Indian Security Policy

WRITTEN UNDER THE AUSPICES OF
THE CENTER FOR INTERNATIONAL AND STRATEGIC AFFAIRS
UNIVERSITY OF CALIFORNIA, LOS ANGELES

A list of other Center publications appears at the back of this book

RAJU G. C. THOMAS

Indian Security Policy

PRINCETON UNIVERSITY PRESS
PRINCETON, NEW JERSEY

Published by Princeton University Press, 41 William Street,
Princeton, New Jersey 08540
In the United Kingdom: Princeton University Press, Guildford, Surrey

Library of Congress Cataloging in Publication Data will be
found on the last printed page of this book

ISBN 0–691–07724–X

Publication of this book has been aided by the Whitney Darrow Fund
of Princeton University Press

This book has been composed in Linotron Baskerville

Clothbound editions of Princeton University Press books
are printed on acid-free paper,
and binding materials are chosen for strength and durability

Printed in the United States of America
by Princeton University Press
Princeton, New Jersey

To
Curian Thomas

CONTENTS

TABLES

TABLES

FOREWORD

INDIA is poorly understood by most Americans. Many of our impressions rest on "exotic" imagery from popular films that creates a general sense of romanticism and poverty. Few Americans think seriously about India in terms of international power and security.

With a per capita income of about $260 per year, India ranks among the twenty poorest countries in the world. But this can be misleading. India is the world's second largest country and largest democracy. The Indian middle class alone is considerably larger than the entire population of Great Britain.

The size of India's population is relatively familiar, but how many Americans realize that India ranks among the ten greatest industrial powers in the world? Or that India has the third largest pool of scientists and engineers in the world after the United States and the Soviet Union? Or that India's nearly 6 percent annual rate of economic growth over the past decade has been approximately double that of the United States? Or that India exports grain and arms as well as industrial products? And that a number of multinational corporations have their headquarters in India?

India's geography and size alone make it a factor to consider in terms of international security. It has the fourth largest army in the world. Both its economic growth and its increasing technological strength enhance its potential power. Understanding Indian security policies is essential for understanding regional stability in South Asia and the Indian Ocean area today. And as long as it is able to combine stable democratic government and economic growth, India is likely to cast an even wider shadow of influence in the future.

Born in India and now an American citizen, Raju Thomas is able to describe India's security policies with a combination of intimate knowledge and scholarly detachment that is all too rare in the field. He gives a thorough account of the evolution of Indian policy from Independence through the assassination of Prime Minister Indira Gandhi and the accession to power of her son, Rajiv. Both the external strategic environment and such domestic factors as the role of the army in internal security, national security planning and decision making, and the evolution of the economic and technological basis for defense decisions are covered.

Thomas describes the influence of the four wars in which India has been involved, and how the division of Pakistan in 1971 left India dominant in the subcontinent. He shows how India's external security policy has evolved from the perception of a limited to a more extended strategic environment. Through its regional entanglements with Pakistan and China, India has been increasingly drawn into Persian Gulf and Southeast Asian affairs. Of particular interest also is Thomas's study of the Indian political system and the nature of Indian domestic politics within which national security decisions are made. He discusses the growing importance of internal security issues in the domestic political debate and the consequent merging of the decision-making bodies that deal with external and internal security factors.

India is currently confronted by a number of difficult defense questions. With Soviet intervention in Afghanistan and American rapprochement with China, India has a delicate set of options in playing off the superpowers. India also faces hard choices in the nuclear area. It conducted an atomic test in 1974, but decided not to follow through with the development of a nuclear arsenal. With Pakistan's increasing nuclear program, India must now decide whether it can continue to avoid a nuclear arms race in South Asia while managing its complex military relationship with China and Pakistan. And, of course, these regional choices further complicate the relations between India and the two superpowers.

In short, Americans cannot afford to continue to think of India in terms of Hollywood images. We must think more clearly about India's domestic and regional politics and the government's security policies. This book is an excellent place to start.

JOSEPH S. NYE, JR.
Harvard University

ACKNOWLEDGMENTS

THIS PROJECT was begun in 1980–1981 on a Ford Foundation grant at the Center for Science and International Affairs, Harvard University, and was continued in 1982–1983 on a fellowship at the Center for International and Strategic Affairs at the University of California, Los Angeles, supplemented by a grant from the Institute for the Study of World Politics. Supplementary funds were also received at various times from the Graduate School of Marquette University and from the Office of Arms Control, Disarmament and International Security, University of Illinois at Urbana-Champaign.

I would like to thank Enid C. B. Schoettle, Program Officer of the International Division of the Ford Foundation, and Kenneth W. Thompson, Director of the Institute for the Study of World Politics and Professor of Government at the University of Virginia, for their assistance. I also thank Paul Doty, Director of the Center for Science and International Affairs at Harvard University; Michael Intriligator, Director of the Center for International and Strategic Affairs at UCLA; Stephen P. Cohen, Co-Director of the Office of Arms Control, Disarmament and International Security at the University of Illinois; John K. C. Oh, former Dean of the Graduate School at Marquette University; and John R. Johannes, Chairman of the Department of Political Science at Marquette University, for their support and assistance in this venture.

I have benefited considerably from the readings and corrections of the manuscript by Stephen P. Cohen; our discussions over the past decade of South Asian security issues have been a constant source of intellectual inspiration. Others who read all or part of the manuscript and provided me with many useful suggestions are Walter K. Andersen of the United States Department of State, P. R. Chari of the Indian Administrative Service, Robert L. Hardgrave, Jr., of the University of Texas at Austin, and Jean-Alphonse Bernard of the French Civil Service. To these gentlemen also I wish to express my gratitude.

This project could not have been completed without the cooperation of several people in India who granted me interviews or other forms of assistance while I did the research in New Delhi. I mention here only a partial list of names: Admirals S. M. Nanda, A. K. Chatterji, and S. N. Kohli; Air Marshals Arjan Singh and P. C. Lal; Generals J. N. Chaudhuri, A. M. Vohra and D. K. Palit; Colonels Rama

Rao and Pyarelal; former Defense Secretary K. B. Lal; former Defence Minister Jagjivan Ram; defense analysts K. Subrahmanyam, T. T. Poulose, P. K. Namboodiri, and Sreedhar; and Bharat Karnad, assistant editor of the *Hindustan Times*. There are many others in the Ministries of Defence and External Affairs whom I interviewed, and I thank them also. In addition I am grateful to the relevant authorities at the Institute for Defence Studies and Analyses, the United Services Institution of India, the Indian Council of World Affairs, Jawaharlal Nehru University, and the Lok Sabha Secretariat for allowing me to use their library facilities. Several graduate research assistants have helped me over the years at Marquette University. They include Anita Bhatia, Sung Joon Roh, and Kathyrn Schalow. The manuscript was meticulously and comprehensively edited by Elizabeth Gretz.

The research in India would not have been possible without the hospitality and assistance provided by my sister and brother-in-law, Susan and Raj Gupta. Finally, I would like to thank my wife Suzanne for her help with various parts of this project and for her forbearance while I continued the research and writing to its conclusion.

RAJU G. C. THOMAS
December 1985

Indian Security Policy

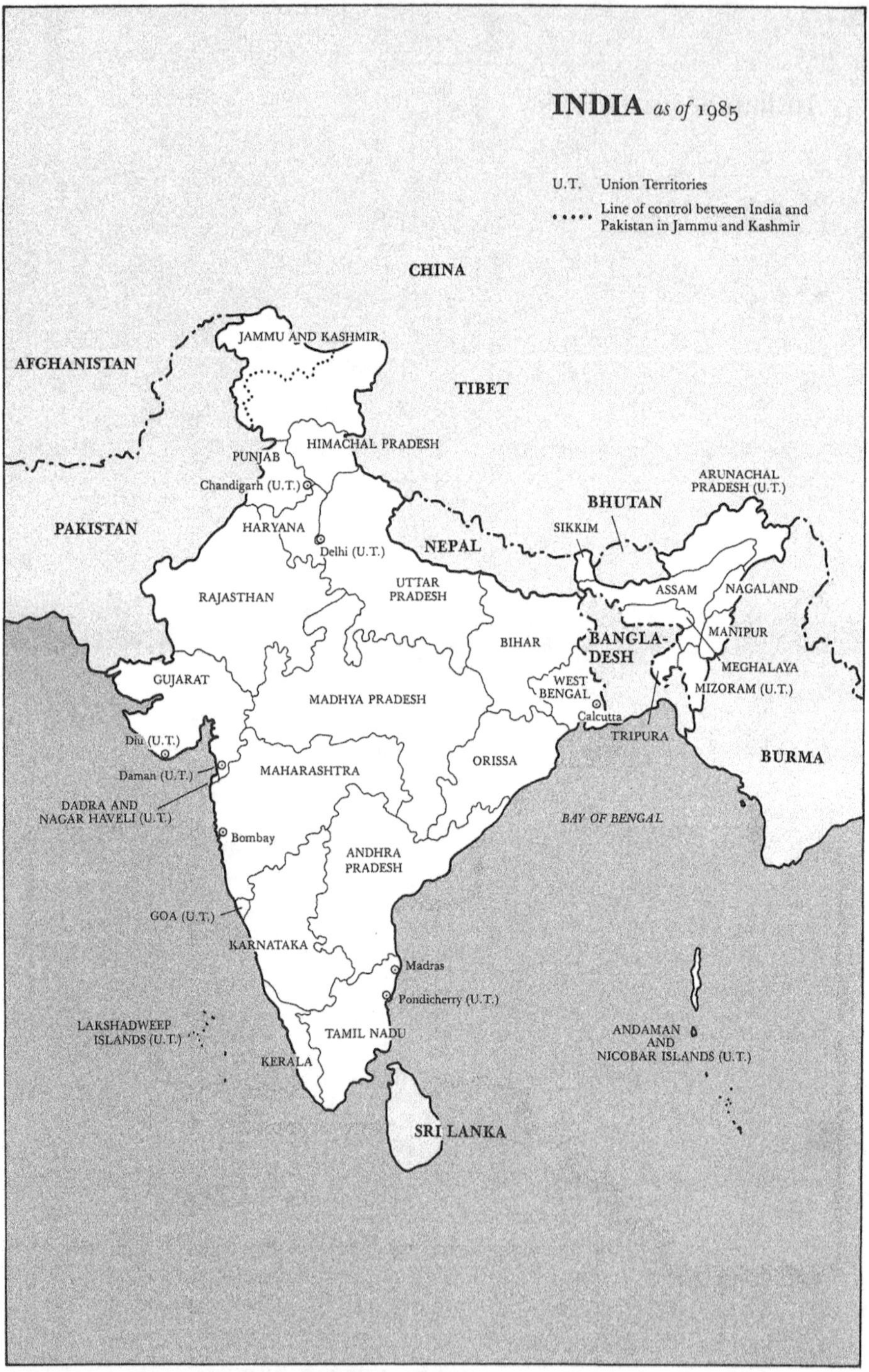
INDIA as of 1985
U.T. Union Territories
Line of control between India and Pakistan in Jammu and Kashmir
CHINA
AFGHANISTAN
JAMMU AND KASHMIR
TIBET
HIMACHAL PRADESH
PUNJAB
Chandigarh (U.T.)
ARUNACHAL PRADESH (U.T.)
BHUTAN
PAKISTAN
HARYANA
SIKKIM
NEPAL
Delhi (U.T.)
UTTAR PRADESH
ASSAM
NAGALAND
RAJASTHAN
BIHAR
BANGLA-DESH
MANIPUR
MEGHALAYA
GUJARAT
WEST BENGAL
MIZORAM (U.T.)
MADHYA PRADESH
Calcutta
TRIPURA
Diu (U.T.)
BURMA
Daman (U.T.)
MAHARASHTRA
ORISSA
DADRA AND NAGAR HAVELI (U.T.)
BAY OF BENGAL
Bombay
ANDHRA PRADESH
GOA (U.T.)
KARNATAKA
Madras
Pondicherry (U.T.)
LAKSHADWEEP ISLANDS (U.T.)
TAMIL NADU
ANDAMAN AND NICOBAR ISLANDS (U.T.)
KERALA
SRI LANKA

CHAPTER ONE

The Dimensions of Indian Security Policy

IN RECENT YEARS, the concept of national security policy has been subject to new interpretations in India. The traditional Indian view—one that is also familiar elsewhere in the world—held that the objective of national security was to protect the state's domestic physical assets, social and cultural values, and life styles from outside aggression, and that security was achieved when the nation possessed sufficient defense to ward off or withstand external attacks.[1] These issues evolve from international politics on both the regional and the global level. Since achieving independence in 1947, the government of India has sought in varying degrees to build up its armed forces and weapons systems to meet the external threats it perceived. These efforts have led to the creation of elaborate decision-making systems at home and the promotion of various policies to procure the weapons needed from domestic sources and abroad.

Since the mid-1970s, however, a second, equally important concept of national security has received greater attention in India. Especially from the standpoint of the newer states in Africa and Asia, security is also perceived to be the ability of a state to avoid, resolve, or ward off internal disorder and violence, secessionist movements, and perhaps even the fragmentation and collapse of the state itself.[2]

[1] According to Morton Berkowitz and P. G. Bock, "national security can be most fruitfully defined as the ability of a nation to protect its internal values from external threats" (cited in Charles F. Hermann, "Defining National Security," in *American Defense Policy*, 5th ed., ed. John F. Reichart and Steven R. Sturm [Baltimore: Johns Hopkins University Press, 1982], p. 19). For an excellent discussion of the problems of defining the external threat, see Arnold Wolfers's chapter entitled "National Security as an Ambiguous Symbol," in his *Discord and Collaboration: Essays on International Politics* (Baltimore: Johns Hopkins University Press, 1965), pp. 147–65. An important and perceptive study of the internal and external worlds of military policy may be found in ch. 1 of Samuel P. Huntington, *The Common Defense: Strategic Programs in National Politics* (New York: Columbia University Press, 1961), pp. 1–24.

[2] The internal component of threats to national security has been recognized by several authors, although problems stemming from internal threats have usually been treated separately in academic writing.

Problems here usually arise from domestic political, economic, ethnic, or social issues.

Given the nature of these domestic questions, it is difficult to evaluate the state of India's internal security. Government policy toward internal "threats" is likely to generate greater domestic controversy than its foreign policy. Indeed, in many Third World countries, state resistance to armed insurrection and other government efforts to maintain the domestic political status quo may not be acceptable to the majority of the population or to powerful segments of it when this implies the perpetuation of an unjust political, economic, and social order. Under these circumstances, internal revolutions against corrupt or unpopular regimes may be perceived as a struggle for the political and economic well-being of the majority. Internal security and long-term stability may then be achieved only through a change in the status quo.

The resort to force against the prevailing government need not always have the support of most of the population. Often such insurrections are conducted by ethnic minorities, a problem that is perpetually faced by the multi-ethnic Indian state. On the other hand, frequent domestic strife may be perceived by others as a constant threat to the nation's security, and they may believe that such problems may be solved only through counterforce actions by the government against the rebellious groups. The Sikh crisis in the Punjab and the Indian government's use of force in 1984 to resolve the issue is one of several examples that have occurred in India since 1947.

Although external and internal security issues are substantially different in nature, they often interact or are interrelated. The problems of domestic insecurity may stem from external political, economic, or military pressures. Thus, for instance, international ideological rivalries may generate domestic strife among opposing factions as well. The foreign and defense policy attitudes of the split Indian Communist parties in the 1960s generally paralleled the factionalization of the international Communist movement and, in particular, reflected the Sino-Soviet split and the policies of these two Communist giants toward the subcontinent and the world. Or again, the economic policies of other nations may be the major cause of internal economic crises that produce civil disorder and violence, as was the case in India during the international oil crisis. Similarly, the buildup of military forces to deal with external threats may be the root cause of military-led coups and the establishment of military regimes at home. An example on the subcontinent is the Pakistani military buildup against India and the usurpation of power from the civilian authorities in Pakistan by the Pakistan army. Conversely, internal conflict may spill over national boundaries and embroil other nations, thus aggravating problems of external security. A good illustration is the East Pakistan revolt and creation of Bangladesh following the Indo-Pakistani war.

As in most western-type democracies, the adoption of defense and weapons procurement policies in India takes place amidst considerable domestic debate. Controversies inevitably arise because the definition of external threat is inherently ambiguous and conjectural. Only after an attack takes place can a state determine whether its estimate of the threat was accurate, and at that point, the state is faced with conditions of war rather than various levels of perceived threats. Thus, in the 1950s, there was a tendency in India to underestimate the Chinese threat from the north and focus almost entirely on the Pakistani threat from the west. This was understandable since, except for some minor skirmishes among local rulers along the border and a British campaign into Tibet, no war had ever taken place between the empires of China and India in recorded history. The birth of India and Pakistan out of British India, on the other hand, had begun with war between the two new states. The Indian perception was also influenced by the fact that the cost of meeting the threat from a more powerful China implied assuming a major defense burden at a time when development needs were perceived to be paramount. Moreover, it was feared that any proposed Indian arms buildup against China would aggravate the situation rather than alleviate it.

The policy maker for Indian security is therefore faced with a dilemma. The underestimation of a threat and the lower defense preparations that result could lead to the erosion of national security if an attack does take place. On the other hand, the exaggeration of a threat and the correspondingly higher defense preparations could lead to countermeasures by the adversary states, which would also lead to a higher threat environment and to the erosion of national security. Political dilemmas may also occur, because any defense policy that is adopted affects the resources distributed to the armed forces and thus alters their size and power. Moreover, the allocation of resources for defense invariably affects civilian programs for social welfare and development. Whether policy involves the determination of external threats, the composition of the defense decision-making bodies, or the allocation of resources to (and among) the armed forces, there are domestic organizations, groups, and individuals in India that stand to gain or lose.

On a broader level, the study of national security policy calls attention to the economic and political development of the state, because this contributes to the successful or unsuccessful management of external and internal security problems. From the standpoint of external security, political stability and economic strength are necessary or desired preconditions for deterring or defending against the enemy. A nation torn with internal strife or facing chronic economic crises cannot deal effectively with external threats. This was demonstrated in 1971 when the Bengali rebellion and Indian intervention culminated in the breakup of Pakistan. From the standpoint of internal security, balanced economic growth and development may allow—at least potentially—a better distribution of material benefits and thereby avoid one important source of conflict. Likewise, the development of workable governmental systems and processes permits more effective channeling of political differences, thus minimizing domestic conflict that may undermine the stability or survival of the state.

Ultimately, a nation's ability to ensure its security against forces from without and to conduct an independent foreign policy depends on the degree to which it is self-reliant in defense. This implies that the country's level of technological capability or access to external technology must be sufficient for it to develop or acquire the weapons it needs. The concept of national security in India has consequently been linked to a strategy of technological self-reliance on a broad scale in the civilian sector, and technological progress in the automotive, aeronautics, electronics, shipbuilding, construction and civil engineering industries has indeed led to a large measure of self-reliance in defense.

These general observations are intended to demonstrate the scope of Indian security policy. This study deals with India's efforts to resolve or cope with external and internal security problems. The strategic, political, economic, and technological issues involved are examined in the broader context of the polity, including the nature of the Indian political and bureaucratic structures and processes; the economic ideologies that underlie planning and the prevailing state of the Indian economy; and the technological strategy adopted and level of capability achieved.

In particular, the significance of this study arises from the substantial changes that have occurred since 1971 in India's external security and domestic political environments. Externally, there have been changes in the strategic relationships of the three great powers—the United States, the Soviet Union, and China. The earlier hostility of both the Soviet Union and the United States toward China (one of India's two traditional adversaries) has been displaced by closer Sino-American ties that are intended to offset the growth of Soviet military power.

More specifically, the Soviet invasion of Afghanistan in December 1979 and efforts by the United States to counter it by renewing military ties with Pakistan (India's other traditional adversary) have complicated India's traditional defense posture; the consequent new round in the arms race between India and Pakistan demonstrates once again the link between great-power politics and the regional politics of the subcontinent. The Indian detonation of a nuclear device in 1974 and Pakistan's efforts to acquire a nuclear weapons capability have similar relevance for both regional and global politics. Such changes in strategic relationships have produced a shift in India's defense strategy from one of "sufficient defense" to thwart Pakistani and Chinese attacks to one of "limited deterrence," implying a broadly based conventional and, if necessary, nuclear weapons capability.

At the domestic level, there have been changes in the Indian concept of national security, both in the decision-making structure and process and in the nature of interservice politics. The concept of national security, as discussed above, has been broadened to include the problems of internal violence and political turmoil. Threats from without as well as violent political strife and armed separatist movements within the nation have resulted in the "merging" of the decision-making bodies that deal with internal and external security. This new structure has altered the previous balance of civil-military relations by reducing the input of the armed forces in the policy process. At the same time, internal violence has increased the use of the armed forces within the country and raised anew the question of the role of the military services in the maintenance of internal security.

The changing external strategic environment has also affected decisions on the distribution of forces and resources among the military services. This has reshaped the politics among the military bureaucracies and the politics between the military services on the one hand and the defense and finance ministries on the other.

The examination of the Indian effort to deal with security concerns on both internal and domestic levels must include a consideration of capabilities—political, economic, and technological. Politically, the issue revolves around India's ability to deal with problems of both external and internal security within its democratic system of government. Especially since the declaration of the Emergency and the suspension of the constitution between June 1975 and March 1977, there have been fears that as problems of security escalated, there would be a greater resort by the government to curtail the rights of citizens and to strengthen those clauses of the constitution that give the government greater powers to enforce law and order and to require individual economic "sacrifice" for the sake of national security.

From an economic standpoint, the question is whether the government is able to raise the necessary resources for defense programs without adversely affecting India's development programs. The consequences of defense allocations on economic growth and price inflation have been questioned. Does more defense mean less development, or can both goals be achieved simultaneously? Varying perspectives exist, and India's defense policy is influenced as much by the domestic economic debate as by the external strategic environment.

As noted earlier, technological capability is a basic determinant of the degree of self-reliance India can achieve in national defense. It is interesting to note that the growth of India's military capabilities is a response both to perceived external threats and to the growth of India's technological manpower. To a certain extent India's technological capabilities determine the country's weapons procurement policies. For example, development of the nuclear energy and space programs, which are principally intended for nonmilitary purposes, has also provided India with potential nuclear weapons and missile system capabilities. For the most part, however, the government has followed a deliberate strategy of acquiring the defense technology needed through indigenous design and development or through licensed production in collaboration with an overseas manufacturer.

The many interrelated components of Indian security policy are discussed in the six chapters that follow. Chapter two concerns assessments of the Indian strategic environment and the conduct of India's foreign and defense policies abroad; chapter three deals with problems of internal security, and the instruments and application of force to resolve threats from within; chapter four studies the nature of the Indian domestic political system within which security policies and programs are formulated and implemented; chapter five examines Indian bureaucratic politics with special reference to the relationships among the Ministries of Defence and Finance and the service headquarters of the army, air force, and navy; chapter six looks at defense issues that arise from the state of the Indian economy and the nature of Indian economic planning and policies; and chapter seven focuses on India's technological strategy and the weapons procurement policies that can be adopted within the constraints of Indian technological capabilities. Finally, a concluding chapter reviews India's security policies as well as the implications for U.S. security policy in the region in the context of the end of Indira Gandhi's rule in November 1984 and the succession of her son, Rajiv Gandhi, as prime minister.

CHAPTER TWO

The Strategic Environment and Defense Policies

NATIONAL security policy primarily addresses the external strategic environment. In the case of India, the external threat environment has undergone revolutionary and evolutionary change at both the regional and global levels. Regional strategic transformations occurred in the Pakistani arms buildup under the SEATO and CENTO defense pacts in the early 1950s, the unexpected war with China in 1962, and the breakup of Pakistan in December 1971 following the East Pakistani rebellion and Indo-Pakistani war. In addition, the explosion of an atomic weapon by China in 1964, the buildup of Chinese nuclear weapons and delivery capabilities, and the detonation of an atomic device by India in May 1974 raised questions about India's nuclear policy and the relationship between conventional and nuclear defense strategies.

A major strategic transformation on the global level of relevance to Indian security policy was the Sino-American rapprochement initiated by President Richard Nixon and Secretary of State Henry Kissinger in 1971. The new ties between the United States and China called for a reevaluation of the threat from the north, especially in the context of the prevailing Sino-Pakistani and Indo-Soviet relationships. Similarly, the Soviet invasion of Afghanistan dramatically affected both global and regional politics by scuttling the Soviet-American strategic arms limitation talks and triggering a new round in the arms race between India and Pakistan.

Changes affecting Indian defense policy in a more evolutionary and distant fashion have occurred in the Middle East, where there has been a massive infusion of sophisticated arms from the Soviet Union and the West; in Southeast Asia following the American military withdrawal and the assertion of Vietnamese military power in the region; and in the Indian Ocean, where the movement of the two superpowers has generated considerable controversy among the littoral states. The changing strategic environment calls attention to both the underlying basis and development of India's external security policy and the nature of the external threat facing India's policy makers.

1. Strategy and Doctrine

The formulation and execution of external security policies raise two central questions. First, what is the nature of the Indian strategic environment and the patterns of security relationships that prevail at the regional and global levels? Clearly, relatively weaker states facing threats at the regional level may seek to draw in the great powers in order to offset the overwhelming threats posed by their neighbors. Conversely, great-power politics and confrontations at the global level may motivate these states to seek regional alliances to further their global objectives. Both reactions produce a web of relationships that link regional and global politics and challenge the ingenuity of the policy maker. Second, what are the defense doctrines and strategies adopted by India to resolve or cope with threats it perceives in its external environment? Defense doctrines are usually developed by major states in order to lend coherence to their overall strategy and to give notice to their adversaries of the resolve and willingness of the nation to carry out appropriate security measures if threatened. In establishing doctine, however, the policy maker also needs to avoid any dogmatic application of policy, which may aggravate rather than alleviate the threat.

Strategic Relationships

Since Independence in 1947, Indian defense policy has pivoted around the prevailing security relationships within the important regional strategic triangle of India, Pakistan, and China.[1] Although India's relations with these states have varied from intense hostility in the past to more cordial relations in recent times, the basic structure of the triangular security relationships at the regional level has not changed. Geographical contiguity together with religious and ideological differences between India and the other two states suggest that varying degrees of security fears are likely to persist in the region even if older issues such as the Kashmir and Himalayan border disputes are eventually reconciled or forgotten. Newer issues in the extended strategic environment that includes the Middle East, Central and Southeast Asia, and the Indian Ocean merely accen-

[1] See my chapter entitled "India: Balancing Great Power Intrusions and Regional Security Interests," in *The Great Power Triangle and Asian Security*, ed. Raju G. C. Thomas (Lexington, Mass.: Lexington Books, 1983), pp. 65–81.

TABLE 2.1
Strategic Alignment in Southern Asia

A. Conflict Postures within or Relevant to Southern Asia

Regional	*Global*
IND–PAK	USA–USSR
IND–PRC	USA–PRC (before 1971)
PAK–AFG	PRC–USSR (after 1963)
PRC–VTM	

B. Conflict Postures and Alignment Tendencies

Interrelated Conflict Postures		*Alignment Tendencies*
PAK–IND & IND–PRC	therefore	PAK + PRC
IND–PRC & PRC–USSR		IND + USSR
AFG–PAK & PAK–IND		AFG + IND
USA–USSR & USSR–PRC		USA + PRC
BNG–IND & IND–PAK		BNG + PAK
BNG–IND & IND–PRC		BNG + PRC
VTM–PRC & PRC–USSR		VTM + USSR
VTM–PRC & PRC–IND		VTM + IND

C. Group Alignment Tendencies

Group 1: PAK, PRC, USA, BNG
Group 2: IND, USSR, AFG, VTM

tuate this basic relationship rather than transform it. Similarly, the regional strategic triangle continues to be linked to the global strategic relationships among the United States, the Soviet Union, and China. Chinese and American efforts to contain the growth of Soviet military power have usually produced military ties with Pakistan, which in turn has tended to increase India's military dependence on the Soviet Union.

Although some of the basic relationships within the regional triangle and between this triangle and other states have been slow to develop or continue to remain blurred, in general these connections have followed the classic theorem of the Indian strategist Kautilya, of the fourth century B.C., regarding the nature of interstate conflict and alliance relationships: "An enemy of my enemy is my friend." Bearing this in mind, it is useful to examine the strategic alignment postures of Southern Asia in summary form (see table 2.1).[2]

Within the regional triangle, the Indo-Pakistani and Sino-Indian

[2] A more detailed study of how this principle has worked in the region may be found in Raju G. C. Thomas, "Security Relationships in Southern Asia: Differences in the Indian and American Perspectives," *Asian Survey* 21, no. 7 (July 1981):689–709.

confrontations have brought about Sino-Pakistani military collaboration. Outside the triangle, the Sino-Soviet and Sino-Indian confrontations have produced Soviet-Indian and Sino-Pakistani military links and in the 1970s the Soviet-American and Sino-Soviet confrontations produced growing Sino-American military ties. Similar political alignments, though not military relationships, have arisen between India and Afghanistan because of their separate confrontations with Pakistan and between India and Vietnam because of their separate but similar conflicts with China. Likewise, the Soviet Union and Vietnam have perceived a common Chinese threat, which has led to the alliance of two communist states against a third. At a more latent stage is the likely development of political and eventually perhaps military ties between Bangladesh on the one hand and China and Pakistan on the other, because of Dacca's growing fear of Indian hegemony. These confrontations and linkages have suggested the common security interests among the Soviet Union, India, Afghanistan, and Vietnam on the one side; and among the United States, Pakistan, and China on the other.

Some alignment tendencies are indirect, as in the case of India's sympathetic policies toward Afghanistan and Vietnam without military links with either state. These ties chiefly derive from the triangular relationship among India, China, and the Soviet Union. Pakistan's past and present military links with the United States have also tended to be based on indirect interests that suggest an underlying quid pro quo arrangement. Pakistan, for instance, has needed American military support to counter first a more powerful India and later a Soviet-backed India, although the Indian military dependency on the Soviet Union was, in turn, partly the result of American arms transfers to Pakistan. Conversely, Pakistan is a useful regional ally in America's global confrontation with the Soviet Union, especially since the United States has been unable to draw India into its anti-Soviet alliance network. In contrast, Indo-Soviet military ties arise more directly from their common perceptions of a Chinese threat.

These interlocking regional and global relationships have tended to constrain the relationships of these states with others in the Middle East or in Southeast Asia, although this may be evident only in a very weak sense. India's ability to forge close ties with Saudi Arabia or Iran under the Shah has been limited in the past by these nations' much closer security ties with Pakistan. Similarly, India's relations with Indonesia declined during the latter part of Sukarno's regime when Indonesia supported Pakistan, particularly at the height of the 1965 Indo-Pakistani war.

Apart from the general pattern of security relationships, there is also a more specific pattern of nuclear or potentially nuclear relationships. Basically, this revolves around Indian perceptions of a Chinese nuclear threat, although the Chinese nuclear arms buildup is in response to the Soviet buildup, which is itself a response to the American buildup. However, any Indian attempt to respond to the perceived Chinese threat with a nuclear weapons program could trigger a Pakistani nuclear weapons program. Theoretically, unraveling this nuclear proliferation chain would require an end of the Soviet-American nuclear arms race, since China is unlikely to terminate its nuclear weapons buildup unless the Soviet nuclear threat is eliminated.[3] Here lies the basis of the Indian argument for equating vertical nuclear proliferation with horizontal nuclear proliferation. Stemming from this is India's refusal to sign the Nonproliferation treaty until the existing "haves" terminate their nuclear weapons programs.

Strategic Doctrines

Changes in the Indian strategic environment over the last thirty years have produced significant shifts in Indian defense policies. However, these changes have rarely, if ever, been officially assessed and communicated through strategic doctrines. There has been, for instance, no Nehru, Gandhi, or Desai defense doctrine similar to an Eisenhower, Nixon, or Brezhnev doctrine. Instead, Indian defense policy has usually been characterized by flexibility and ambivalence, although elements of doctrine may be discerned in foreign and defense policy actions from time to time. Perhaps the continuing Indian adherence to nonalignment and the rejection of military alliances as the basis for national security may be considered a security doctrine.[4] The initial phase of nonalignment carried with it a corollary set of principles based on the Buddhist philosophy of *panch-*

[3] See Raju G. C. Thomas, "The Nuclear Club and Affirmative Action," *The Bulletin of the Atomic Scientists* 37, no. 5 (May 1981):60–62. For a collection of Indian views on India's nuclear policies, see T. T. Poulose, ed., *Perspectives of India's Nuclear Policies*, (New Delhi: Young Asia Publications, 1978); and K. Subrahmanyam, ed., *Nuclear Myths and Realities: India's Dilemma* (New Delhi: ABC Publishing House, 1981). A detailed study is provided in Ashok Kapur, *India's Nuclear Option: Atomic Diplomacy and Decision-Making* (New York: Praeger Publishers, 1976).

[4] See Raju G. C. Thomas, "Nonalignment and Indian Security: Nehru's Rationale and Legacy," *Journal of Strategic Studies* 2, no. 2 (September 1979):153–71.

shil, or the "Five Principles of Peaceful Coexistence," which was intended to maintain peace and security between India and China.

There are two possible interpretations of the Indian policies of nonalignment and *panchshil.* On the one hand, they suggest primarily a foreign policy philosophy on which there exists a broad national consensus rather than doctrines intended for specific security purposes. The philosophical interpretation suggests that such policies are inherently superior—politically and even morally—to the alternative of military alliances and confrontations that may lead to war. According to this perspective, however, nonalignment and *panchshil* may not prevent wars, although they may be made less likely. Thus in a speech to the Indian Council of World Affairs in 1949, Jawaharlal Nehru stated that "if war comes, it comes. It has to be faced. The prevention of war may include providing for our own defence and you can understand that, but that should not include challenges, counter-challenges, mutual cursing, threats, etc. These certainly will not prevent war, but will only make it come nearer."[5] On the other hand, such policies may be viewed as the best method of preventing wars and maximizing Indian security. For instance, Nehru concluded that military weakness did not necessarily imply diminished security or political influence abroad. Instead, greater security could be achieved through noninvolvement in military blocs. The latter interpretation, which perceives a different purpose in the policies though not a different substance, suggests an Indian security doctrine. As will be discussed, the first attitude was implicit in Nehru's policy before the 1962 Sino-Indian war, and the second after it.

Whether or not the declared foreign policy of nonalignment is conceived as an Indian doctrine, its practice has reflected widely varying interpretations over time. The practice of nonalignment has ranged from nonparticipation in the cold war under Prime Minister Nehru in the 1950s, to military alignment (implicit especially in the signing of the Indo-Soviet treaty) under Prime Minister Indira Gandhi at the height of the 1971 Bangladesh crisis, and to neutralism in the late 1970s under Prime Minister Morarji Desai.[6] The

[5] Jawaharlal Nehru, *India's Foreign Policy: Selected Speeches, September 1946–April 1961* (New Delhi: Publications Division, Government of India, 1961), p. 24.

[6] For two Indian interpretations of India's nonalignment policy, see K. P. Misra, "The Concept of Nonalignment: Its Implications and Recent Trends," in *Studies in Indian Foreign Policy,* ed. K. P. Misra: (New Delhi: Vikas Publications, 1969), pp. 90–106, and P. R. Chari, "Nonalignment and International Security," in *Role of Nonalignment in a Changing World,* proceedings of a seminar held at the India International

practice of nonalignment, therefore, suggests a flexible standard to be adapted to the prevailing strategic circumstances. In a sense, it is not strategic doctrine that has produced certain foreign and defense policies under different Indian prime ministers and governments; rather, a variety of policies seems to suggest adherence to some vague concept of a doctrine broadly termed nonalignment.

It has also been debated whether Indian prime ministers have adhered to the strategic principles declared in the mandala (a circle within a center of a nucleus) by Kautilya in his *Arthashastra* (Statecraft).[7] This geopolitical treatise held that neighboring states should be treated as enemies or potential enemies, and states beyond the neighboring states as friends, since these according to the mandala theory would be the enemies of the neighboring states. In the Indian case this would imply that Pakistan, China, Bangladesh, Burma, and Sri Lanka are enemy or potentially enemy states; Iran, Afghanistan, the Soviet Union, Japan, and the other states of East and Southeast Asia would be potential allies. But Indian leaders do not appear to have based foreign policy on this theory although, as we saw earlier, the regional and global strategic relationships appear to conform to the Kautilyan principle, "an enemy of my enemy is my friend." Thus, rather than treat China as an enemy state, the government under Nehru persistently sought the friendship and goodwill of the Chinese Communist leaders as the cornerstone of Indian foreign policy and the prerequisite for maintaining peace in Asia. Subsequently, war with China in 1962 seemed to have confirmed the theory of Kautilya, at least with regard to some of the key states of Asia. India's principal neighbors—Pakistan and China—proved to be enemy states, while some of the neighbors of these neighbors—Afghanistan, the Soviet Union, and Vietnam—turned out to be friendly states whose policies against China or Pakistan were useful in drawing the Chinese or Pakistani threat away from India. Nevertheless, Indian foreign policy, especially from the mid-1970s on, has again rejected Kautilyan principles in seeking to cul-

---

Center, April 28–29, 1976 (New Delhi: India International Center Publications, 1976), pp. 68–81.

[7] This mandala theory of Kautilya was seen as implicit in Indian foreign policy by Brig. Noor Hussain, the director of the Institute of Strategic Studies in Islamabad, Pakistan. See his paper entitled "Indian Regional Foreign Policy: Strategic and Security Dimensions," presented at the International Conference on Security Futures, University of Illinois at Urbana. May 7–8, 1984. An excellent study of Kautilya's "statecraft" may be found in George Modelski, "Kautilya: Foreign Policy and International System in the Third World," *American Political Science Review* 58, no. 3 (September 1964): 549–60.

tivate the friendship of China and even Pakistan, and to foster regional cooperation in South Asia. Although accepting the tendency of states to conform to the mandala theory, India seeks at the same time to neutralize or eliminate this tendency as the basis for guaranteeing its security. Kautilyan principles appear to be accepted in theory but rejected in practice.

Some ambivalence may also be observed in the Indian rejection of "balance of power" politics in principle as undesirable and dangerous legacies of western strategic thought. This has not, however, led to its rejection in practice. Indeed, one of the earliest proponents of balance-of-power politics was Kautilya, and Indian policy makers, especially under the Indira Gandhi administration appear to have studied and practiced the Kautilyan art with subtlety and finesse. For example, India's policies in the Middle East and toward the superpowers have often adroitly offset the power plays of its two main adversaries, Pakistan and China. Pakistani efforts to forge Islamic military alliances in the Middle East to deal with the question of Kashmir have been countered by skillful Indian diplomatic moves in the region; and India has maneuvered successfully between the two superpowers to neutralize the military threat perceived to arise from the growth of Chinese military capabilities. In both cases, tacit "alignment" strategies, regionally and globally, have served India as well as if it had entered into formal alliances with one or more of these states. Here Kautilyan principles are rejected officially but practiced unofficially.

Whether or not strategic doctrines may be discerned from the policies and actions of Indian prime ministers, India's basic military postures tend to be more explicitly defined. Since the wars with Pakistan in 1948 and 1965 and with China in 1962, the underlying Indian military posture has been to maintain a slight edge in military capabilities against Pakistan and to maintain minimum border defense capabilities for a holding operation against China until some form of superpower intervention could be sought to forestall further advance. In other words, until recently, India's defense posture has for the most part followed a policy of sufficient defense, which itself has varied from little or no defense (as toward China in the 1950s) to minimum defense.

The policy of virtually no defense capabilities directed toward China was justified by Nehru on three grounds: the futility of conducting a major arms race against a much larger country with greater resources; the potentially destabilizing effects of such an arms race on economic development at home; and the more idealistic belief that adequate security could be obtained through diplomatic "policies" rather than through the raising of "armies." This mix of beliefs was reflected in the twin pillars of Nehru's foreign policy, *panchshil* and nonalignment. *Panchshil* was first enunciated at the 1953 Bandung conference of nonaligned nations and then embodied in the preamble of the 1954 Sino-Indian treaty on Tibet, which enjoined both states to follow peaceful methods to resolve all disputes.[8] When *panchshil* appeared to have failed in the war of 1962, a policy of minimum defense against China and matching military capabilities against Pakistan was followed until the late 1970s.

In the 1980s, the policy of sufficient defense appears to be moving toward one of limited deterrence. This posture is based on a two-fold approach. First, the continuation of recent efforts to improve relations with Pakistan and China is expected to lower their motivation to settle unresolved disputes through force. Second, the continued expansion of Indian military capabilities is expected to make the resolution of unsettled disputes by the use of force less worthwhile. The strategy of limited deterrence thus attempts to achieve a deterrent Indian military capability and to reduce the incentives among potential adversaries to resort to force.

However there appears to be a contradiction in the present Indian defense policy of reducing regional tensions with its traditional adversaries (and thereby also the threats perceived), while also increasing Indian military capabilities. As discussed in chapter seven on technology, the explanation for this may be found in part in the technological momentum in India that makes defense programs that appear feasible also appear necessary. However, the more fundamental explanation may be found in a major shift in India's perception of its strategic environment. From a previous view of a limited strategic environment that essentially called for military responses to threats stemming from the growth of Pakistani and Chinese military capabilities, newer conditions beyond the subcontinent in the Middle East, Central and Southeast Asia, and the Indian Ocean have begun to indicate an extended strategic environment.

This newer perspective has links with the traditional strategic environment that centered around Pakistan and China. India's grow-

[8] *Panchshil* called for (a) mutual respect for each other's territorial integrity and sovereignty; (b) mutual nonaggression; (c) mutual noninterference in each other's internal affairs; (d) equality and mutual benefit; and (e) peaceful coexistence. See Nehru, *India's Foreign Policy: Selected Speeches, September 1946–April 1961*, p. 99.

ing economic ties with the oil-exporting Islamic nations in the Middle East and the concurrent growth of Pakistani military links with these states have tended to bring about the strategic interdependence of the South Asian and Middle Eastern regions.[9] The prospect that a Sino-American quasi-military alliance may develop in the future in response to the growth of Soviet military power, the continuation of Soviet-American naval rivalry in the Indian Ocean and accompanying efforts to seek military bases among the littoral states, and the potential U.S.-assisted rearmament of Pakistan and China in response to the Sovet invasion of Afghanistan suggest the need to monitor developments beyond the traditional security perimeter of the Indian subcontinent. Thus the continued buildup of Indian defense capabilities may be rationalized in terms of the extended strategic environment that has continued to generate arms buildups in Pakistan and China despite improving relations among India, Pakistan, and China.

II. The Traditional Strategic Framework

Indian defense planning has been primarily based on the wars with Pakistan and China. After the first Indo-Pakistani war, subsequent defense programs in India were essentially responses to Pakistani defense programs. Likewise, the war with China produced defense preparations in India to meet the recurrence of what were expected to be similar wars along the Himalayan borders. Two more wars with Pakistan in 1965 and 1971, accompanied by various threats by China to intervene, firmly established a dual Sino-Pakistani threat. Thereafter, contingency defense planning was based on what American strategists might have termed a "one-and-a-half war" principle.[10] India was prepared to fight a full-scale war with Pakistan on all levels—ground, air, and sea; against China, conventional defense preparations were based on the assumption that Sino-Indian wars would follow the pattern of the 1962 conflict. The third major contingency that appeared to call for an Indian response was the growth of Chinese nuclear capabilities.

[9] M. G. Weinbaum and Gautam Sen, "Pakistan Enters the Middle East," *Orbis* 22, no. 3 (Fall 1978):595–612.

[10] This term was used to indicate a shift in American global strategy under the Nixon doctrine. The one-and-a-half war principle involved maintaining the capabilities to fight a major war and a minor war concurrently as opposed to the earlier policy of being able to fight two major wars and a minor war.

The Pakistani Threat

Except for a brief spell after the 1962 Sino-Indian war when the Chinese threat suddenly seemed awesome, the principal threat to India from Independence on has been perceived to come from Pakistan. In 1947, as India and Pakistan emerged as independent states out of British India, the separation of Pakistan from India to provide a homeland for Indian Muslims brought with it the potential for international conflict between two sovereign states over issues that were previously confined to the domestic level. The partitioning of the subcontinent also added new disputes based on territorial claims, the sharing of river waters, the treatment of the substantial religious minorities left on both sides, and the division of assets between the two states. These newer issues merely fueled preexisting Hindu-Muslim antagonisms. Foremost among post-Independence areas of conflict was the dispute over Kashmir; two of the three Indo-Pakistani wars since Independence have been fought over this previously autonomous state, whose population was predominantly Muslim but ruled by a Hindu maharajah.

The security problem in the 1980s revolving around the Pakistani threat is not fundamentally different. For Pakistan, the Kashmir issue cannot be considered resolved as long as India occupies and rules the more desirable two-thirds of the state. Although the intensity of the Pakistani claim on Kashmir has mellowed since the separation of Bangladesh from Pakistan as a result of the third Indo-Pakistani war in 1971, from the Indian standpoint, the basic threat from Pakistan will remain as long as the possibility exists that Pakistani leaders might be tempted to seize Kashmir whenever an advantageous military situation arises.

Meanwhile, Pakistan, as the smaller and weaker state, sees India as a major threat. Although the perception of this threat arises partly from Pakistan's unwillingness to accept the territorial status quo in Kashmir, Pakistani fears stem more broadly from the traditional belief that the Indian government has never fully accepted the creation of Pakistan and would seek opportunities to reabsorb it into a greater India. This has produced an effort in Pakistan to match or more than match Indian military capabilities, which has further aggravated the threat perceived by India. As in all such rough balance-of-power strategies, there is a Pakistani tendency to err on the high side in the procurement of high quality weapons from abroad, consequently producing an equal or greater Indian counterreaction. The conventional arms race on the subcontinent displays the classic security dilemma.

There are other factors contributing to the threats and counterthreats perceived by the two subcontinental neighbors. External arms races have become interlocked with the Indo-Pakistani arms race. The politics of the cold war and the attendant arms race between the United States and the Soviet Union, as well as the later concurrent Sino-Soviet struggle for security, have produced efforts by the United States, the Soviet Union, and China to draw India and Pakistan into their security plans. This has resulted in the infusion of arms from these states to the subcontinent and the upgrading of Indian and Pakistani capabilities. Most of Pakistan's arms have been obtained from the United States and China; India's have come from the Soviet Union. Note that unlike West European arms transfers to the region, which have been primarily motivated by profit considerations, the arms transfer policies of the three major powers have been based on their strategic interests.

From India's point of view, arms supplied by the United States to Pakistan in the late 1950s and early 1960s under the anticommunist alliance networks—the South-East Asia Treaty Organization (SEATO) and the Central Treaty Organization (CENTO)—greatly aggravated Indian security. The transfer of American arms under these two pacts included M-47/48 Patton tanks, and F-104 Starfighter, F-86 Sabre, and B-57 Canberra combat aircraft. The total value of such equipment at the time was estimated at about $700 million; to this was added another $1.3 billion worth of infrastructural support such as communications equipment systems and training programs.[11] India responded to the Pakistani acquisition with purchases from Great Britain and France—Centurion and AMX-13 tanks, and Hunter MK-56, Mystère-IVA, and B-1 Canberra combat aircraft.[12]

It is noteworthy though not unexpected that Indian and Pakistani defense programs have tended to be mirror images of each other.

[11] These figures were obtained from testimony given by Townsend Hoopes, principal deputy assistant secretary of defense for international security in 1966, in hearings before the Senate Subcommittee on Near Eastern and South Asian Affairs. See U.S. Senate, Committee on Foreign Relations, *Arms Sales to the Near East and South Asian Countries*, 19th Cong., 1st sess. (Washington, D.C., Government Printing Office, 1967).

[12] For a general discussion of these problems, see Raju G. C. Thomas, *The Defence of India: A Budgetary Perspective of Strategy and Politics* (New Delhi: Macmillan Co. of India, 1978; Columbia, Mo.: South Asia Books, 1978), pp. 155–56, 176–79, and 214–17.

This results in part from the traditional Indian response to the Pakistani threat, which was based on the principle of "matching capabilities" through the 1950s and much of the 1960s. After the 1971 Indo-Pakistani war the principle of matching capabilities shifted briefly to one of maintaining a "slight edge" and subsequently, as discussed below, India's military posture moved to one of clear superiority over Pakistan.

Thus, during the year of the 1965 Indo-Pakistani war, India matched Pakistan's ground forces.[13] There were seven Indian infantry divisions to Pakistan's six, and one armored division and one armored brigade on either side. Although there was a minor advantage in the number of infantry soldiers on the Indian side, this was adequately offset by the superior striking power of Pakistan's armored units, equipped with American-supplied tanks, as compared to India's armored units, equipped with British and French tanks. Similarly, on the eve of the 1971 Indo-Pakistani war, there were thirteen Indian infantry divisions to Pakistan's twelve, with two armored divisions on either side, although the second Indian armored division had not been fully constituted when war broke out.

In the air, the principle of matching capabilities tended to be more qualitative than quantitative. In 1965, despite an Indian numerical superiority in combat aircraft of about two to one, Pakistan's American-supplied planes were qualitatively better than the British and French aircraft purchased by India. A similar situation prevailed in 1971 except that India's new policy of maintaining a "slight edge" was beginning to show. Pakistan's aircraft were becoming obsolete as a result of the arms embargo to the subcontinent imposed by the Johnson administration after the 1965 war. Meanwhile, India's MiG-21 interceptors produced in India with Soviet technical assistance, the indigenous HF-24 Marut fighter-bomber, and Indian purchases of Soviet Sukhoi-7B fighter-bombers provided the Indian air force with both numerical and some marginal qualitative superiority over the Pakistani air force. The exception to all of this was the Pakistani purchase of Mirage-IIIs from France, against which India had no qualitative match.

At sea the situation was somewhat different. Both navies had been neglected until the 1965 war, after which there were steady Indian purchases of various kinds of Soviet vessels such as the F-class submarines, Osa-class missile boats, and Petya-class frigates. In

[13] These figures are drawn from *The Military Balance, 1965–66* and *1971–72* (London: International Institute for Strategic Studies, 1965 and 1971).

Pakistan, however, the navy continued to be neglected and the consequences were severely felt during the 1971 war, when hostilities were extended to sea for the first time since Independence. Nevertheless, it is important to note that even here the difference in force levels was insignificant. The reason that the Indian navy was able to overwhelm the Pakistani navy was the comparatively lower degree of importance attached to the development of naval power by Pakistani decision makers under conditions in which both navies had been generally neglected relative to the other two services.

The situation of only slight superiority may not be readily apparent, given the Indian military victory over Pakistan in the 1971 war. Note, however, that Indian victory was rapid only in East Pakistan, where Pakistan had three demoralized infantry divisions fighting without air cover, compared to India's three infantry and three more mountain divisions that had been diverted from the China border to invade the eastern Pakistani province. In addition the Indians were greatly aided by the local revolutionary forces, the Mukti Bahini, which accelerated the Pakistani defeat in the east. On the western front, the forces were evenly matched and the outcome was indecisive, although India overall may have gained some minor territorial advantage, especially in the less-defended desert areas of Pakistan's Sind province.

The shift from matching capabilities to slight edge became more apparent after the 1971 war. Both the Indian and Pakistani armies improved the teeth-to-tail ratio of their infantry divisions and reconstituted them in smaller numbers. Therefore, without much change in total numbers, there were eighteen infantry divisions and two armored divisions on either side by the mid-1970s. With the further obsolescence of Pakistan's American and Chinese equipment and a corresponding improvement in Indian strength with the acquisition of Soviet T-54/55 and PT-76 tanks and Indian-made Vijayanta tanks, India began to gain the advantage militarily.

Similar trends may be seen in the air and at sea. Although obsolescence had begun to overtake the Indian air force's Hunters, Canberras, Sukhois, Maruts, and Ajeets, the Anglo-French Jaguar fighter-bomber was acquired to provide India with a decisive qualitative edge over Pakistan's front-line combat aircraft, the French Mirage-IIIs and -Vs. At sea, despite the procurement of Agosta- and Daphne-class submarines as well as Spruance-class destroyers, there was not much comparative improvement in the Pakistani navy. Further Indian acquisitions of Soviet naval vessels, especially the first generation Kashin-class destroyers and Nanuchka-class corvettes, began to provide India with clear naval superiority.

By the mid-1970s, the traditional Pakistani threat in the air had become much less menacing as their capabilities declined and Indian capabilities rose. Given the significant difference in the resources, manpower, and gross national product of the two sides, the Indian objective of maintaining a slight edge could be achieved without serious strain on the Indian economy. The transfer of American weapons to Pakistan two decades ago may have aggravated the Pakistani threat and compelled the government of India to dig into its foreign exchange reserves in order to make offsetting purchases from Britain and France. But this was not beyond Indian capabilities at that time, when large amounts of foreign exchange had been accumulated in India as a result of the international trade boom in primary commodities after the Korean War.

Nevertheless, meeting the threat from Pakistan still posed two major problems. The first involved the ability of Pakistan to gain access periodically to qualitatively superior weapons, as in their procurement of advanced American weapons in the 1950s and 1980s. Although India may have the resources to respond to such developments, it cannot always match the performance capabilities of such acquisitions. India's purchases of British and French Hunters, Mystères, and Centurions in the 1950s cannot be considered an equivalent response to Pakistan's Starfighters, Sabres, or Pattons obtained from the United States. Indian decisions in 1981 to procure the British Jaguar and Soviet MiG-23 aircraft and the Soviet T-72 tank have since been "leap-frogged" by Pakistan's ability to obtain American F-16 fighters, M-901 TOW antitank missiles, and the upgunned and improved version of the M-48 tank, the M-48A5. Needless to say, this has produced an Indian effort to overtake Pakistani acquisitions through an order for French Mirage-2000s and attempts to obtain American Harpoon missiles and the Phalanx antiaircraft guns.

The second problem concerned the vulnerability of Kashmir to a surprise Pakistani attack, particularly if Pakistan had qualitatively superior weapons. Although in a war of attrition a Pakistani invasion to seize Kashmir might be eventually rolled back, there is little chance that this could be accomplished in the kind of two- to three-week Indo-Pakistani wars that occurred in 1965 and 1971. Under these circumstances, whether India adopts a policy of matching capabilities, slight edge, or even clear superiority, Pakistan still constitutes a serious and credible threat. Hence India strongly objected in

1981 to the transfer of American weapons to Pakistan that are considered technologically superior to those of India, even though India has numerical superiority in weapons.

The Chinese Threat

Indian perceptions of the Chinese threat have varied. In the early 1950s, there were misgivings in the Lok Sabha (the lower house of the Indian Parliament), especially among opposition party members, about the Chinese claim to Tibet. After the Indian decision in 1954 to sign an agreement with China that granted Beijing full rights over Tibet, the leader of the Praja Socialist party, Acharya Kripalani, criticized the *panchshil* provisions of the treaty and claimed that the agreement amounted to "a surrender of Indian rights in Tibet."[14] Similarly, members of the Jan Sangh party in 1953 had warned the government of the danger from the north. They too disapproved of the next year's *panchshil* agreement and later denounced the Chinese suppression of the 1959 Tibetan revolt. That revolt and the founding of the procapitalist and anticommunist Swatantra party in 1959 brought more warnings of the Chinese threat.

Such warnings were sporadic before the 1962 war and were not taken seriously by the Congress government of Nehru. Much of the time there appeared to be general agreement in both government and opposition political circles with Nehru's "panchshil" approach to China. Nehru's approach to security based on nonalignment and peaceful coexistence appeared to be a practical solution that neutralized a potential Chinese threat.[15] The alternatives of a major

[14] See *Lok Sabha Debates*, 2d ser., vol. 5 (May 5–21, 1954), col. 7548; and vol. 18 (August 19, 1958), col. 1676; and 3d ser., vols. 10–11 (December 1962), col. 5134. See also K. Raman Pillai, *India's Foreign Policy: Basic Issues and Political Attitudes* (Meerut, Uttar Pradesh: Meenakshi Prakashan, 1969), pp. 80–84 and 137–41.

[15] During the debate in the Lok Sabha in 1954 on the Sino-Indian treaty, Nehru tried to justify the agreement. His arguments revolved essentially around the strategic inevitability of the situation. He declared that it was a matter of importance to both India and China, which now had 1,800 miles of common frontier, that they live in peace, friendship, and with respect for each other's sovereignty and territorial integrity. By doing so a large part of Asia would be guaranteed peace. "Several Honourable Members have referred to the 'melancholy chapter of Tibet.' I really do not understand. I have given the most earnest thought to this matter. What did any Honourable Member of this House expect us to do in regard to Tibet at any time? Did we fail, or did we do a wrong thing? The fact is, and it is a major fact of the middle of the twentieth century, that China has become a great power, united and strong." See Nehru, *India's Foreign Policy: Selected Speeches, September 1946–April 1961*, p. 304.

arms race with China seemed self-defeating if it risked crippling India's economic development efforts.

The 1962 Sino-Indian war suddenly changed these beliefs. The Himalayan ranges were no longer a protective barrier and economic development without adequate defense was now seen to be futile. The immediate Indian reaction to the shock of military defeat was to seek arms from the West, especially from the United States and Britain. Indeed, in the months following the war, there were some suggestions that India might consider joining the network of western military alliances. In July 1963, the government of India decided to accept an Anglo-American offer of joint air exercises later that year to test the country's air defense. Although the announcement also stated that the western air squadrons would leave India soon after the conclusion of the exercises and that nonalignment continued to be the government's official policy, Indian military dependence at that time on the West, especially for air cover against China, became evident.[16]

The establishment of military ties with the West proved unworkable. Pakistan objected to any major western arms transfers to India on the grounds that such a policy would cause an erosion of Pakistani security. Thereafter, as the Chinese threat receded, both India and the western powers became less interested in entering into a military arrangement. The 1965 Indo-Pakistani war ended the limited American military assistance to India, which had consisted mainly of an offer to build an ammunition factory and to transfer some light mountain guns and transport planes.

The net result of these events and negotiations was that the Chinese threat by the late 1960s was perceived to be secondary and capable of being met through indigenous weapons production. Since the Sino-Indian border controversy had subsided and both sides seemed relatively satisfied with their lines of actual control, the Chinese threat was viewed as arising indirectly from China's quasi-alliance relationship with Pakistan. The 1965 and 1971 Indo-Pakistani wars indicated that barring Indian attempts to assert claims to territory now occupied by China, the government in Beijing was only likely to threaten limited military operations during such wars in order to relieve the military burden on its ally Pakistan.[17] The In-

[16] See *Times of India*, July 13, 1963.

[17] Gen. J. N. Chaudhuri, chief of army staff during the 1965 Indo-Pakistani war, discusses the potential Chinese threat and how to meet it in *India's Problems of National Security in the Seventies* (New Delhi: United Services Institution of India, 1973), pp.

dian belief that Chinese military involvement on the subcontinent would remain limited was reinforced by perception of an overwhelming Chinese fear and preoccupation with the Soviet threat, which would prevent it from deep involvement in a military crisis where it had limited interests.

Under these circumstances, defense against China did not envisage anything beyond a replay of the 1962 Sino-Indian war. Planning has therefore not changed substantially since the early postwar plans were drawn up in 1963. These called for raising ten mountain divisions equipped with light arms and transport equipment and the substantial upgrading of communication lines and other infrastructural support systems along the border areas. Almost all of these objectives were achieved within seven years and have been maintained ever since.

The optimistic official Indian assessments of the Chinese threat appear to have been proved correct. Yet it must be noted that during both the 1965 and 1971 Indo-Pakistani wars, China threatened to intervene. The manner and level of such an intervention is not clear. India has assumed that it would have followed the same pattern as in the 1962 war against which, at least by 1971, India was much better prepared. However, it is unlikely that a new Chinese military thrust would be similar to the 1962 campaign if no further advantage could be gained. Whatever China's political or technological reasons for not escalating the war to involve air combat in 1962, these restraints would presumably have been overcome in a second Sino-Indian war. The emphasis of a new Chinese military thrust might also shift from the Northeast Frontier Agency (NEFA, now renamed Arunachal Pradesh), where Indian troops were routed in 1962, to the Ladakh sector in Kashmir. This would enable China to coordinate a joint Sino-Pakistani military operation against India effectively.

An Indian military response to such a scenario has probably been considered at the Indian armed services headquarters. But there are few indications that Indian military plans in the 1960s and 1970s were based on dealing with a two-front war. Indeed, in 1971 during the Indo-Pakistani war, defense measures against China were reduced; three mountain divisions from the NEFA sector were diverted to join the invading Indian forces in East Pakistan and at least one mountain division from the Ladakh sector was diverted to join the

21–26. See also Lt. Gen. A. M. Vohra's more specific assessment in *China's Strategic Posture in the 1980s* (New Delhi: United Services Institution of India, 1972).

Indian forces deployed along the West Pakistan borders. Defense against China in 1971 appeared to rest on the calculation that China would have sympathy for the revolutionary liberation movement in Bangladesh and on the deterrent value of the newly signed Indo-Soviet treaty.

The Nuclear Threat

The first serious discussion of a nuclear threat to India and the need for an Indian nuclear military response occurred after the Chinese atomic test of 1964. Between the first Chinese atomic test and the Indo-Pakistani war of 1971, considerable debate took place in India on whether the nation should develop its own nuclear weapons. The Chinese military ultimatum to India during the Indo-Pakistani war of 1965 and the continuation of Chinese atomic tests thereafter reinforced the case of the Indian pro-bomb lobby. Its basic argument was that even if current Chinese nuclear weapons development was directed against the Soviet Union, there was no guarantee that China would not resort to veiled nuclear blackmail during times of crisis on the subcontinent.[18] Particularly given the unexpectedness of the 1962 Sino-Indian war, proponents of the bomb suggested that it would be safer to take the nuclear contingency into account in Indian defense preparations. According to this argument, ten mountain divisions would prove futile if the conventional threat from the north were to escalate to a nuclear level in future Sino-Indian confrontations.

However, the lobby—led in Parliament mainly by the Jan Sangh party at the time—did not find many adherents. A nuclear India was considered more likely to enhance the Chinese nuclear threat rather than to reduce it. In addition, the growing intensity of the Sino-Soviet rift and the continuing hostility of the United States toward the Communist government in Beijing implied that one or both superpowers would restrain China from resorting to nuclear threats against India. Above all, the buildup of nuclear weapons in India would carry no defense or deterrent purpose when India lacked the delivery capabilities to attack Chinese nuclear launching systems (a counterforce capability) or Chinese industrial and population centers (a countervalue capability). The Indian space program had barely begun in late 1963 and was still a long way from developing its own rocket systems that could be converted into a ballistic missile delivery system.

[18] See Chaudhuri, *India's Problems of National Security*, pp. 16–20.

The case for building nuclear weapons was rejected for the time being by the government. Some who sought the rejection of those proposals went to the other extreme, advocating the renunciation of nuclear weapons for moral reasons. Proponents were found among those who considered themselves the disciples of Mahatma Gandhi; one of the more important figures was Morarji Desai, finance minister under both Nehru's and Indira Gandhi's Congress governments and later prime minister in the Janata government. Although after the 1962 Sino-Indian war Desai had claimed that India must make every sacrifice for the defense of the country, his subsequent attitude toward nuclear weapons—especially during his term as prime minister—suggested a preference for unilateral renunciation.[19] The rationale for renunciation was based on the nature of weapons that destroyed not only present generations but also future ones.

India's nuclear policy during the years before the Bangladesh crisis of 1971 lay between these two extremes and may best be termed a strategy of maintaining the nuclear weapons option. This policy implied that India would not make nuclear weapons but would constantly threaten to do so. Such a posture served two purposes. First, given the commitment of the two superpowers to stop the spread of nuclear weapons, both presumably would be anxious to provide India with credible nuclear guarantees against the Chinese nuclear threat in order to prevent India from carrying out its veiled threats to become a nuclear power. Second, the Indian threat in a more general sense would presumably pressure the existing nuclear "haves" into reducing and eventually eliminating their existing nuclear stockpiles.

Although the first purpose solved the more immediate problem of the Chinese nuclear threat, the second was expected more distantly to improve the general climate of international security, which would benefit India as well. Thus, by wielding the nuclear weapons option, India avoided the vulnerability that existed in a unilateral renunciation of nuclear weapons on the one hand and the increased security dangers that would arise from embarking on a nuclear arms race with China on the other. And indeed, this ap-

[19] Morarji Desai reiterated this again in 1981, two years after the Janata government fell from power, when he called upon Prime Minister Indira Gandhi to renounce nuclear weapons. Ironically, Desai's preference for the unilateral renunciation of nuclear weapons had been rejected earlier by the majority of his colleagues in the Janata government. See *New York Times*, June 3, 1981.

proach proved to be a useful policy between the extremes of inaction and action.

III. The Extended Strategic Framework

A series of events and emerging conditions in the early 1970s had begun to transform the strategic environment of India. At the subcontinental level, the breakup of Pakistan and the creation of Bangladesh altered the basic power relationship between India and Pakistan. The reduction of Pakistan to half its former size suggested a reduction in its threat to India. At the global level, Sino-American rapprochement carried with it the prospect of military collaboration between the United States and China, which appeared to increase the Chinese conventional and nuclear threats to India. Also on the global level, Soviet-American naval rivalry in the Indian Ocean affected regional military balances.

At the regional level three developments were relevant to India. First, the various arms races in the Middle East, made possible by the abundance of petrodollars in the region, had indirect influence on India because of Pakistan's military links with some of these states. Second, Vietnam's military campaigns in Southeast Asia coupled with China's military involvement in the region was of concern for India, although to a lesser extent than Middle East conditions. Finally, India's detonation of a nuclear device in 1974 followed by Pakistan's scramble to acquire a nuclear weapons capability involved the substantial alteration of the nuclear threat to India and a revision in India's nuclear policy.

More than ten years later, Indian responses to these developments, some of which were still evolving, had not reached beyond the stage of national debate. Defense policies during the 1975–1985 decade were primarily based on the strategic conditions that prevailed in the year following the breakup of Pakistan. At that time the situation was referred to as the "new realities of the subcontinent," implying that India had established regional predominance and its own sphere of influence. India had to contend with a defeated and demoralized Pakistani military regime; the new state of Bangladesh was considered to be pro-India and without much military capability. The failure of China to intervene or indulge in diversionary military tactics during the 1971 Indo-Pakistani war indicated that the traditional Chinese threat had not substantially altered.

Nevertheless, from the mid-1970s on, the Ministry of Defence, especially under the earlier Congress government of Indira Gandhi, together with members of the public, began to question defense programs based on the traditional strategic framework.[20] Events beyond the traditional perimeters of Indian defense—realignments in the great power triangle, regional events in the Middle East and Southeast Asia, and the new nuclear threat—were now considered to have a bearing on Indian security. Note, however, that the perceived extended strategic environment has not displaced the traditional strategic environment. The new perceptions tended, rather, to broaden and complicate the older security concerns.

The Changing Global Strategic Perspective

As the 1970s began, there were two nascent conditions that raised doubts about the adequacy of India's traditional defense posture and programs. The first was the diplomatic rapprochement between the United States and China initiated in 1971 by the Nixon-Kissinger administration; the second was the movement of the superpower navies into the Indian Ocean in the early 1970s.

Sino-American normalization was aided by President Yahya Khan of Pakistan. The close ties between Pakistan and China enabled the Pakistani government to arrange for the secret diplomatic visit of Kissinger to Beijing in July 1971. This was also the year of the prolonged Bangladesh crisis. At a time when millions of Bengali refugees were flowing into the volatile Indian states of West Bengal, Assam, and Tripura to escape Pakistani military oppression, the Nixon administration's policy was to avoid alienating Pakistan, the ally of China, so that U.S. efforts to normalize relations with China would not be jeopardized. Indeed, Kissinger threatened not to dissuade China from intervening in the subcontinent on Pakistan's behalf should India resort to force to solve the Bangladesh issue.[21] The complex experiences of 1971 led India to fear the development of a U.S.-China-Pakistan security axis.

India perceived U.S. policy as assigning greater importance to Sino-American relations than to the problems of the subcontinent. On the other hand, the place of India in Soviet global strategy was high, especially with the intensification of the Sino-Soviet rift. Thus

[20] See, e.g., *Report: 1974–75* (New Delhi: Ministry of Defence, Government of India), p. 2. For a 1979 assessment of India's defense needs for the 1980s, see the chapters by P. R. Chari, B. Chakravorty, Bhabani Sen Gupta, Ravi Kaul, P. M. Mehrotra, Vishal Singh, K. R. Singh, R.V.R. Chandrashekhar Rao, T. T. Poulose, and Bhupinder Singh in *Strategic Environment in South Asia during the 1980s*, ed. D. D. Khanna (Calcutta: Naya Prokash, 1979).

[21] Reported in the *Hindustan Times*, August 8, 1971.

when the United States and China decided to end their confrontation and establish close diplomatic relations—during the height of the East Pakistani rebellion—India found the time ripe for a closer relationship with the Soviet Union. The Indo-Soviet Treaty of Peace and Friendship signed in August 1971 was the Indian response to the regional crisis in Bangladesh and global developments between China and the United States. According to the traditional interpretation, the immediate purpose of the treaty was to raise doubts in the minds of the Communist leadership in Beijing about possible Soviet reactions to any Chinese attempt to engage in diversionary tactics along the Himalayan borders. Especially for India, the treaty was seen as a deterrent to American or Chinese intervention in the event of an Indo-Pakistani war.

However, an alternative interpretation suggests that China had informed Pakistan that in the event of an Indo-Pakistani war it would not intervene.[22] Moreover, according to this perspective, India had already received intelligence reports regarding China's communications with Pakistan. Under these circumstances, the signing of the Indo-Soviet treaty had little to do with the impending war on the subcontinent but more to do with Indian apprehension about Soviet relations with Pakistan. The problem with this alternative (which generally endorses the official Indian position that the treaty had little to do with neutralizing the Chinese threat during the 1971 crisis) is the timing, as well as certain clauses that suggest a deterrent purpose. For instance, Article 9 suggested the possibility of military and diplomatic assistance by the Soviet Union to India under crisis conditions. For the Soviet Union, Article 8 was important because it restrained India from joining an anti-Soviet alliance system and committed New Delhi to strict neutrality in the event of a western-backed Chinese conflict with the Soviets.[23] Since the conditions projected in Article 8 were not about to happen and those projected under Article 9 had already occurred, it would seem that the treaty, from India's perspective, would discourage China from

[22] This interpretation was suggested to me by Walter K. Andersen of the U.S. State Department. He does not think "that the 1971 Indo-Soviet Treaty of Peace and Friendship had a great deal to do with China—from the Indian perspective, that is. Of course, it did from the Soviet perspective. Rather it seems to have had more to do with Indian concerns regarding Soviet relations with Pakistan." This view is spelled out in some detail in a volume in preparation by Richard Sisson and Leo Rose on decision making during the 1971 Bangladesh crisis.

[23] The Indo-Soviet treaty is reproduced in full in the *Current Digest of the Soviet Press* 23, no. 32 (September 7, 1971):5.

risking intervention in a conflict where sympathy for the revolutionary liberation movement in Bangladesh outweighed the temptation to support China's ally Pakistan against India, its one-time antagonist and ally of its arch rival, the Soviet Union.

The new regional and global alignments that occurred during the 1971 crisis subsided over the next few years. Despite some recommendations within the United States to establish a military alliance with China, or at least to bolster Chinese defense capabilities through the transfer of western military technology, Sino-American ties did not advance much beyond a close economic and diplomatic relationship.[24] Every American president from Nixon to Reagan has made a visit to China and declared the need to maintain close ties, but there have been no major military agreements that demonstrate the commonality of strategic interests.

Meanwhile, the dreaded Washington-Islamabad-Beijing axis also proved to be less ominous than first expected by India. India's relations with all three states showed signs of improvement. Likewise, Indo-Soviet relations reverted to the situation that prevailed before 1971; the 1971 treaty did not prove to be the forerunner of a closer Indo-Soviet military alliance. After the Bangladesh issue was settled, India granted no military bases or other military favors to the Soviet Union and Soviet influence over India's foreign policy was hardly visible. Indeed, even during the 1971 Indo-Pakistani war, Soviet ability to control Indian military actions appeared minimal.

The Afghanistan crisis that flared up in December 1979 demonstrated that these potential military relationships had remained latent. The Soviet invasion produced reactions that reemphasized the commonality of security interests among the United States, Pakistan, and China on the one hand, and the Soviet Union and India on the other. Chinese and most American perceptions of Soviet expansionist motives were confirmed and their fears of the growth of Soviet military power were heightened. The need to contain perceived Soviet expansionist plans in the Persian Gulf and South Asia increased the importance of Pakistan and China in Washington's regional military strategy. After various efforts by the Carter admin-

[24] Although an alliance relationship did not come about under the Carter administration, the basis for such a relationship was noted in 1978 by Carter's national security advisor, Zbigniew Brzezinski, who stated in a visit to China: "The basic significance of the trip was to underline the long-term strategic nature of the United States' relationship to China. It is based on certain congruence of fundamental interests, and it reflects a mutual understanding that these interests are enduring" (*New York Times*, May 28, 1978).

istration to develop somewhat informal and low-level military relationships with Pakistan and China, the Reagan administration in June 1981 concluded two major and separate agreements for the transfer of sophisticated American military equipment and technology to these two states.[25] Later, in 1984, President Ronald Reagan agreed to provide China with nuclear reactors for generating nuclear power without formal or written assurances that these reactors would not be diverted to military use. This agreement was formalized in August 1985.[26]

Ironically, the 1981 military agreements came just after the visit of the Indian foreign minister to Pakistan and during the visit of the Chinese foreign minister to India, events that promised closer relations within the regional triangle of adversary and allied states. The Reagan administration's decisions to supply arms to Pakistan and to liberalize the transfer of military technology to China not only set back India's improving relations with its traditional adversaries but also increased Indian military dependence on the Soviet Union.

The second global trend affecting Indian security was the movement of the superpower navies into the Indian Ocean. This movement stemmed from a British decision based on a 1966 Labour party Defence White Paper that declared that British forces were greatly overextended between Europe and the Far East, making forces east of Suez relatively expensive as well as incapable of relieving any serious military crisis.[27] In keeping with this assessment, Britain evacuated Aden by 1968; other forces in the Persian Gulf, Malaysia, and Singapore were drastically cut back by 1979. The legacy of empire was ended in a major step taken in December 1974, when British Defence Minister Roy Mason announced the liquidation of all British military deployments east of Suez with the exception of a small force to be maintained in Hong Kong for internal security purposes.[28]

As the British carried out their withdrawal, the Soviet Union reacted initially by calling in 1969 for a collective security pact in Asia

[25] See *New York Times*, June 5 and 16, 1981.

[26] For an Indian discussion of this agreement, see "Sino–US Nuclear Diplomacy," *Economic and Political Weekly* (Bombay), August 10, 1985, p. 1335.

[27] A discussion of this is found in L. W. Martin, "British Defence Policy: The Long Recessional," Adelphi Papers, no. 61 (London: International Institute for Strategic Studies, November 1969).

[28] See *Times of India*, December 9, 1974; and the *Hindu*, March 28, 1975.

to fill the vacuum caused by the departure of the British.[29] The scheme, apparently directed against China, was rejected by India and other Asian states at the time. Subsequently, both superpowers decided to fill the vacuum and there were efforts to obtain bases and docking facilities in the Indian Ocean. The initial major foothold obtained by the United States was the island of Diego Garcia; the Soviet Union was able to obtain docking facilities in Aden in Southern Yemen.

The growing Soviet-American naval presence led to demands by a number of littoral states for their withdrawal, and a resolution was passed in 1971 in the United Nations General Assembly to maintain the Indian Ocean as a "zone of peace." Such demands, however, were accompanied by reservations or silence on the part of other states, who viewed the displacement of the superpowers in the area as advantageous only to some of the major regional powers such as India, Indonesia, and Iran. Calls for the creation of a "zone of peace" were thus countered by other demands for a "balance of naval forces" that would include the great powers.

The Pakistani position on this issue was revealed in 1974, when Prime Minister Zulfikar Ali Bhutto not only welcomed the setting up of a U.S. naval base at Diego Garcia but also renewed an earlier invitation made to President Nixon to establish an American air and naval base on the Arabian Sea between Karachi and the Iranian frontier.[30] Pakistan additionally tied the concept of the "zone of peace" in the Indian Ocean to the concept of a "nuclear-free zone" in South Asia, since the absence of the superpowers in the Indian Ocean might result in nuclear blackmail by India. The Pakistani proposal was rejected by India unless it was extended universally to

[29] This call was first made by Chairman Leonid Brezhnev in a speech to the International Conference of Communist and Workers' Parties in Moscow on June 7, 1969. Brezhnev stated that the international course of events "places on the agenda the task of creating a system of collective security in Asia" (*Pravda* and *Izvestia*, June 8, 1969, pp. 1–4; published in full in *Current Digest of the Soviet Press* 21, no. 23, July 2, 1969, pp. 3–17).

[30] According to Seymour Weiss, the U.S. State Department's director of the Bureau of Political-Military Affairs, in testimony given before Congress, both Pakistan and China favor the establishment of a U.S. naval base at Diego Garcia. See House of Representatives, Committee on Foreign Affairs, *Proposed Expansion of U.S. Military Facilities in the Indian Ocean*, hearings before the Subcommittee on the Near East and South Asia, February 21 to March 6, 1974 (Washington, D.C.: U.S. Government Printing Office, 1974), pp. 21–49. Prime Minister Bhutto's offer to provide a military base to the United States was indicated in a report by Drew Middleton, "Persian Gulf Emerging as Military Focus," *New York Times*, June 22, 1975.

include all the nuclear powers and especially China.[31] Consequently, despite a call by the Soviet Union in 1977 for the demilitarization of the Indian Ocean and periodic American responses that it was willing to consider various nonaligned summit proposals, prospects for a general solution of any kind on the Indian Ocean remained distant.

The strategic realignment among Moscow, Beijing, and Washington and the naval developments in the Indian Ocean provide the issues at the global level for rethinking and revising Indian defense policies and programs in the 1980s. Events and conditions in the Middle East and Southeast Asia also display the broader strategic interests of India.

The Middle East and Southeast Asia

The global changes in the Indian security environment have been accompanied by changes at the regional level, within the traditional security perimeter. These include the separation of Bangladesh from Pakistan, which has resulted in a more concentrated Pakistani defense system, the new Bangladesh armament program mainly assisted by China, and some fluctuations in China's bilateral relations with the states of South Asia. China has sought to improve its diplomatic relations with India while at the same time it has continued to upgrade its military ties with Pakistan and has embarked on developing similar ties with Bangladesh. These changes suggest the need for some adjustments in India's traditional defense posture; they do not call for radical revisions.

The more important regional changes have taken place further west (the Middle East) and east (Southeast Asia) of India's traditional defense perimeter. These changes provided the basis for the debate in the 1970s over whether India's strategic doctrines and programs need substantial revision. Significantly, the threats that may arise from these two regions have no precedent to guide policy. However, the revised strategic environment of the 1970s and 1980s is directly or indirectly related to the older Pakistani and Chinese threats.

Events that have transformed the Middle East and increased the strategic relevance of this region for India are several and interrelated. The transformation began with the outbreak of the Arab-Israeli war in October 1973 and the ensuing Arab threat to cut off oil to the West. By 1974, the Organization of Petroleum Exporting

[31] *Hindu*, May 31, 1978.

Countries (OPEC) had been consolidated to fix higher oil prices and the Palestinian cause had given way to the promotion of economic gain for the members of OPEC at the cost of both western industrialized and developing nations. The accumulation of petrodollars in the Middle East subsequently led to the large-scale purchase of sophisticated arms by Iran, Iraq, Saudi Arabia, Libya, and some of the Gulf sheikdoms.

The purposes of these autonomous arms buildups were mixed. In almost all cases, they were intended to protect the individual state's oil resources against a variety of regional and global threats. In some instances the reasons arose from the need to present a stronger and more aggressive Arab posture on the Palestinian dispute with Israel; in others buildups were motivated by local territorial disputes as in the case of the Shat al-Arab waterway dispute between Iran and Iraq. The several arms races in the Middle East both aggravated older security fears and added new ones.

None of these arms buildups in the Middle East is presumed to be currently directed against India. The security implications of these trends are more long-term and arise from Pakistan's military ties and involvements in the region, from the strategic interests and arms transfer policies of the great powers in both the Middle East and South Asia, and from various economic dependencies and linkages that have developed between the states of the subcontinent and the oil-exporting Islamic states.

In the Middle East, Indian apprehensions initially arose from the large-scale arms purchases by the Shah of Iran from 1974 until his overthrow in 1979. The Iranian military buildup partially stemmed from the Shah's perceptions of a developing Kabul–New Delhi–Baghdad axis supported by the Soviet Union. By 1978, Moscow had signed treaties of "peace and friendship" with India, Iraq, and Afghanistan. All three bilateral treaties included clauses that called for military consultation with the Soviet Union in the event that one of the signatories was involved in military conflict. The defense programs of all three states were also heavily dependent on the supply of Soviet weapons. Following the separation of Bangladesh from Pakistan in 1978, the Shah also committed himself to defending the territorial integrity of Pakistan and concurrently sought to strengthen the CENTO alliance of which Pakistan and Turkey were a part.[32]

[32] See Amin Saikal, *The Rise and Fall of the Shah* (Princeton: Princeton University Press, 1980), pp. 143–46.

Earlier, the Shah had provided some logistical support for Pakistan during the 1965 Indo-Pakistani war and subsequently transferred a reported ninety F-86 Sabre fighters to Pakistan that took part in the 1971 Indo-Pakistani war.[33] Such relatively modest military support for Pakistan before the oil crisis had little impact on the military balance of the subcontinent but demonstrates the potential implications of large-scale arms accumulation in the Persian Gulf for Indian security. In his major defense plan initiated in 1974, the Shah undertook to spend $28 billion over five years to raise a powerful and well-equipped Iranian military. Allocations to the defense effort during this time averaged between 10 and 17 percent of the Iranian gross national product of about $75 billion.[34] Indian defense allocations were 3 percent of the Indian GNP of approximately $110 billion. Between 1974 and 1978 Iran spent $36 billion on defense compared to about $12 billion spent by India. Iran's shopping list included F-16 and F-14 Tomcat fighters, Airborne Warning and Control Systems (AWACS), Phoenix missiles, *Spruance*-class destroyers, and British Chieftain tanks.[35]

The fall of the Shah brought about an end not only to the Iranian military shopping spree but also to the military collaboration between Pakistan and Iran under CENTO. However, Washington's failed effort to sponsor Iran as a regional military power to act as a policeman of the Gulf has since been replaced by an effort to bolster Saudi Arabian military power. Again, as in Pakistan's CENTO links with the Shah's Iran, similar military links are being established between Pakistan and Saudi Arabia. The underlying basis of this regional military relationship was not a western-sponsored alliance—although the United States encouraged the new Saudi-Pakistani relationship—but the need to promote Islamic defense cooperation. Over the last decade, Pakistan has been providing modest military training programs for the Saudi ground forces. In 1980, such cooperation reportedly led to an agreement that would station two Pakistani army divisions in Saudi Arabia.[36] These divisions were to be financially supported and equipped by the Saudi government.

[33] Saikal, *Rise and Fall*, pp. 156–57.

[34] From *The Military Balance, 1979–80.*

[35] Compiled from *World Armaments and Disarmament SIPRI Yearbook 1980* (New York: Crane Russak and Company, 1980), pp. 98–100.

[36] See *Statesman*, December 24, 1980; and *New York Times*, February 6, 1981. For an assessment of Pakistan's military capabilities and links with the Middle East, see B. Chakravorty, "Pakistan: A Factor in Indian Security in the 1980s," *Strategic Environment in South Asia in the 1980s*, ed. Khanna, pp. 55–61.

Two further compensating divisions were to be raised in Pakistan and to be similarly financed and equipped by Saudi Arabia. Note, however, as of 1985 there is no evidence that such an agreement was in fact carried out. Pakistan does not have more than two brigades stationed in Saudi Arabia as of 1985.

United States efforts to arm both states are part of its Middle East defense strategy against the Soviet Union. However, India saw such moves as the development of a U.S.-Pakistani-Saudi military axis that might eventually prove detrimental to Indian security. Unlike Iranian and Jordanian arms transfers to Pakistan before 1971, Saudi Arabia has not transferred military equipment to Pakistan in the past. However, the potential for military transfers exists although there may be some problems of immediate absorption by the Pakistani armed forces of the technology and the weapons supplied to Saudi Arabia by America. Thus the growth of Saudi Arabian military power has indirect relevance for Indian security, as did the earlier growth of Iranian military power. The main military purchases by Saudi Arabia from the United States between 1975 and 1985 included sixty advanced F-15 Eagle fighter-bombers equipped with fuel tanks for extended ranges beyond 1,000 miles and carrying air-to-air Sidewinder missiles; AWACS and K-135 refueling planes; M-60 A3 tanks, TOW antitank missiles, and Dragon and Maverick missiles.[37] Saudi purchases from France included a number of 2,000-ton frigates equiped with Otomat antiship missiles and Dauphine helicopters equipped with AS-15 TT medium-range missiles.[38]

The Saudi arms buildup and Pakistan's Saudi links were supplemented by other similar but lesser developments elsewhere in the Middle East. In the early 1970s, India was apprehensive about several squadrons of Mirage-IIIs and -Vs in the air forces of Libya and the United Arab Emirates.[39] Pakistani air force personnel were used to train the pilots in these two countries to fly their Mirages, and the transfer of these planes remained a possibility that would have bolstered the existing Pakistani deployments of such planes. Although no Mirages have as yet been transferred to Pakistan, memories of the transfer of American F-104 Starfighter and F-86 Sabre combat

[37] Information drawn from various issues of *The Military Balance*; the *Washington Post*, March 27, 1981; and the *New York Times*, March 2, 1980, and April 22 and June 14, 1981.

[38] *International Herald Tribune*, October 15, 1980.

[39] See Raju G. C. Thomas, "Aircraft for the Indian Force," *Orbis* 24, no. 1 (Spring 1980):85-101.

aircraft to Pakistan in the past produce apprehension among Indian decision makers.

Finally, the prolonged Iran-Iraq war provides lessons on the importance of Middle East security issues for Indian defense planning. Despite the fall of the Shah, with whom India had maintained good relations, India and Iran continued their good relationship with the rise of the Khomeini-led revolutionary government. India has also managed to maintain good relations with the Saddam Hussein regime in Iraq despite the outbreak of the Iran-Iraq war. Pakistan similarly maintains good relations with both states. Pakistan's ties have been closer to Iran than to Iraq; India's ties have been closer to Iraq than to Iran. Thus both India and Iraq have security and military links with the Soviet Union and closer economic relations. Although Iranian-Pakistani military ties ended with the fall of the Shah, close economic, cultural, and political ties remain. Under these circumstances, both India and Pakistan are concerned about the effects of the war on the stability of the Persian Gulf and the South Asian region.

Fears of the spread of Khomeini-directed or -inspired Islamic fundamentalist revolution in the conservative Arab states in the Gulf have led to their financing of the Iraqi war effort. These fears are shared in India, which has a large Muslim minority. One report estimated that Saudi Arabia was paying 60 percent of the cost of Iraq's war effort, Kuwait 30 percent, and the other countries in the Gulf region making up the rest.[40] The complexity of this situation is apparent in that the Saudis, who support Pakistan over India, also support Iraq—which, like India, has a treaty of peace and friendship with the Soviet Union. Saudi financing thus helped Iraq to buy Soviet weapons. On the other hand, Syria and Libya, who also have similar treaties with the Soviet Union, support Iran. The lessons of the Iran-Iraq war suggest the possibility of Arab financing of Pakistani wars against India in the future.

Just as India's strategic interests in the Middle East arise because of Pakistan's political and military links with the region, India's strategic concerns in Southeast Asia arise from China's political and military involvement in this region. The region, however, is of comparatively lesser strategic relevance than the Middle East. Although the earlier American involvement in the Vietnam war was essentially of moral concern to India, subsequent Vietnamese activities in Kam-

[40] See John Kifner, "Unlikely Allies Emerge over Persian Gulf War," *New York Times*, July 17, 1984.

puchea and Laos, the Sino-Vietnamese conflict, and Soviet-Vietnamese military ties make the region of some strategic importance to India. Vietnamese efforts to impose an Indochinese federation to include Laos and Kampuchea under their dominance may not in itself concern Indian defense planners. The strategic linkages between the subcontinent and Indochina tend to be latent and are found in the parallels between Indian and Vietnamese confrontations with China and Indian and Vietnamese ties with the Soviet Union.

Although Sino-Indian ties have improved in recent years and both sides appear content with the de facto line of control over their disputed Himalayan borders, various global and regional alignments suggest the potential for future tensions and conflicts. As noted earlier, the Soviet invasion of Afghanistan strengthened the collaboration among Pakistan, China, and the United States and in turn increased India's military dependence on the Soviet Union. The increasing Indian dependence on the Soviet Union tends to produce a sympathetic response in New Delhi for Soviet-supported Vietnamese policies in Southeast Asia.[41]

Indian interest in military developments in Indochina was highlighted when China resorted to a small strike against Vietnam in February-March 1979. A second border war occurred between China and Vietnam in July 1984. Both wars were in response to Hanoi's military actions in Kampuchea,[42] where in 1978, the Chinese-supported Pol Pot regime had been overthrown and displaced by the Vietnamese-supported Heng Samrin regime. This led to the first Chinese military retaliation against Vietnam to teach the government in Hanoi "a lesson." China subsequently continued to support the guerrilla forces of Pol Pot against the Heng Samrin regime; the second Sino-Vietnamese war was intended to draw Vietnamese forces away from Kampuchea. Not only was the 1979 Chinese invasion of Vietnam reminiscent of its action against India in October 1962, it was also carried out while India's external affairs minister under the Janata government, A. B. Vajpayee, was visiting China. This was the first visit by a high-ranking Indian official to China since 1962; it had the declared purpose of bringing about the normalization of Sino-Indian relations. The invasion of Vietnam

[41] This dependence and commitment was also apparent under the Janata government. See, for instance, the report of Prime Minister Desai's visit to Moscow in the *Times of India*, June 4, 1979.

[42] The second Sino-Vietnamese border war is reported in the *New York Times*, July 14 and 15, 1984.

prompted Vajpayee to call off the remainder of his scheduled visit to China in protest.[43]

The other strategic link may be found in the two bilateral treaties of peace and friendship: the Indo-Soviet treaty of August 1971 and the Soviet-Vietnamese treaty of November 1978. Both include clauses calling for mutual consultations to ensure peace in the region should one of the signatories be attacked. In addition, both India and Vietnam have received most of their military equipment from the Soviet Union. Not unexpectedly, these treaties were entered into when both India and Vietnam were devising military plans to resolve regional crises against states that were the allies of China, Pakistan and Kampuchea.

The Vietnamese military action against Kampuchea in 1978 must have seemed safe because it took place after the signing of the Soviet-Vietnamese treaty. The similar treaty signed by India in 1971 had enabled Indira Gandhi to resolve the East Pakistani refugee and Bangladesh issues through military action. The Indian move had thrown the credibility of the then-existing Sino-Pakistani quasi alliance into doubt. China could not or was not willing, it seemed, to intervene on behalf of an ally, especially when the antagonist had a "peace and friendship" treaty with the Soviet Union. On the other hand, Soviet inaction during the Chinese invasion of Vietnam would conversely raise doubts about the Soviet willingness to intervene on the subcontinent if China should choose to intercede in the future on behalf of its ally Pakistan. Nevertheless, there are also understandable reasons for these failures to intervene by China and the Soviet Union. In 1971 China was concerned by the need to balance its special ties with Pakistan with its general commitment to national liberation movements. Similarly, Soviet inaction in Indochina in 1979 was related to Vietnamese sensitivity about direct Soviet military intervention.

The common military dependence of both India and Vietnam on the Soviet Union, their common confrontations with China, and the prevalence of similar strategic objectives, such as the maintenance of a modicum of regional hegemony in South Asia and Southeast Asia, have produced common political support for each other and economic cooperation between the two countries. A brief exception occurred during the Janata government's rule when India refused

[43] See reports in the *Times of India*, March 10 and 11, 1979. The issue of the Chinese invasion of Vietnam and its relevance for Indian security was subsequently discussed in the Indian parliament. See *Lok Sabha Debates*, 6th ser., vol. 13, no. 39 (April 18, 1978), cols. 305–308 and 315–16; vol. 24, no. 21 (March 20, 1979), cols. 275–84.

to recognize the Heng Samrin regime in Kampuchea following the ouster of the pro-Chinese Pol Pot regime through armed Vietnamese action. The return of Indira Gandhi to power in January 1980 quickly brought about Indian recognition of the Heng Samrin regime, although this government remained unrecognized by the rest of the world. The new Congress government also provided Vietnam and Laos with $75 million in credit for the purchase of Indian industrial and agricultural goods.[44]

Apart from these latent Indian alignments with other states that have Soviet military ties such as Iraq and Vietnam, India's strategic interests in the Middle East and Southeast Asia are also based on existing or anticipated economic ties. In the Middle East such economic ties are being fostered with the oil-producing and -exporting countries. Members of OPEC—especially Iraq and Iran (until war broke out between them in October 1980), Saudi Arabia, the United Arab Emirates, Kuwait, and Libya—are the major sources of Indian oil imports.[45] The rest of India's oil imports have come from the Soviet Union, Mexico, and Venezuela.

In order to offset the rising cost of oil imports, a concerted effort has been made by the government of India to increase the exports of Indian raw materials and industrial manufactures to these countries. The oil-exporting Persian Gulf states have also attracted large numbers of Indian workers and professional personnel. In 1980, for example, this overseas labor force remitted almost $500 million back to India.[46] At the same time, New Delhi has sought to attract Arab investments to India. There has been a similar effort to develop economic ties with the Association of South-East Asian Nations (ASEAN), Thailand, Malaysia, Singapore, Indonesia, and the Philippines. Although India's exports to Indonesia continued to

[44] *Times of India*, December 15, 1980.

[45] Until the Iranian revolution of 1979, 40 to 50 percent of India's oil imports came from Iran and another 20 to 30 percent from Iraq. The war that broke out between these two countries in October 1980 eliminated almost 70 percent of India's oil import sources. Subsequently, new agreements were sought with the Soviet Union, Saudi Arabia, the United Arab Emirates, Libya, Nigeria, Mexico, and Venezuela. Two years later, the Soviet Union and Saudi Arabia provided about half of India's oil imports. See Peter Sinai, *New Lamps for Old* (New Delhi: Indian Documentation Service, 1980); *Hindustan Times*, December 16, 20, and 24, 1980; *Times of India*, January 1, 1981; *Statesman*, January 28, 1981; and *India News* (Washington, D.C.), February 9, 1981.

[46] See Sinai, *New Lamps for Old*, p. 11. For various Indian exports and investment contracts in this region, see the *Economic Times* (Bombay), January 6, 1981; *India News*, February 23 and April 6, 1981; and *Hindustan Times*, December 26, 1980.

rise in the 1980s, efforts to increase trade with other members of ASEAN were hampered by the decision to recognize the Heng Samrin regime.[47]

All of these issues demonstrate the expansion of India's strategic interests to include the Middle East and Southeast Asia. Events and crises in these regions are perceived by the Defence Ministry and Indian defense analysts to have a bearing on Indian defense planning. As in the case of global strategic changes, the Indian recognition of problems at the extended regional level has not yet produced a well-defined defense policy in response. There have been some demands from the Indian navy to expand into a blue water navy from its basically coastal deployment on the grounds that the newer strategic circumstances point to a more important naval role. However, Indian naval programs have continued to be modest. Meanwhile, the programs of the other two services have continued to emphasize modernization rather than revision and expansion. Defense programs in the early 1980s have remained in a holding pattern with most reaction to these extended relationships confined to diplomatic efforts to neutralize potential threats. There is also the belief, especially among the non-Congress opposition parties, that the enlarged defense perspectives may be based on worst-case scenarios and that proportionate defense programs could prove self-defeating if adopted.

One complication in adopting an extended defense posture is that this may unnecessarily produce fear of India in the Middle East and Southeast Asia that has not previously existed. This may then upset India's cordial relations with the oil-producing nations in the Middle East and adversely affect India's oil supplies and external trade relations. Such an extended defense posture could also trigger arms races that cut across the traditional regional sectors of the Middle East, South Asia, and Southeast Asia, eventually posing an enormous defense problem for India. On the other hand, although the net result of an extended Indian defense posture may be a new interregional arms race, the alternatives may be gross Indian military unpreparedness in the event of a wider military conflagration.

The Revised Nuclear Perspective

India's traditional nuclear policy appeared to have been discarded in May 1974, when the "option" was in fact carried out and an un-

[47] For India's exports to the Southeast Asian nations, see the annual issue *India: 1981* (New Delhi: Publications Division, Ministry of Information and Broadcasting, 1981), pp. 339–40.

derground atomic test was conducted at Pokharan in the Rajasthan desert. The nuclear club before India joined it had been confined to the five permanent members of the United Nations Security Council. Although the 1974 Indian atomic test was officially declared to be for peaceful purposes, there were certain underlying strategic and domestic political reasons for the decision. It, in turn, has produced external reactions, especially in Pakistan, that have transformed India's nuclear environment.

The initial impetus for proceeding with plans to conduct an atomic test goes back perhaps to the Bangladesh crisis and the Indo-Pakistani war of 1971. During that prolonged crisis, efforts by President Nixon and Secretary of State Kissinger to seek normalization of relations with China suddenly raised doubts about the credibility of external nuclear guarantees against China. The immediate Indian reaction to these new global developments was to sign the Indo-Soviet treaty of August 1971.

However, doubts remained about the long-term effects of the new Sino-American relationship on India's nuclear security. The double nuclear indemnity provided by both superpowers against China before 1971 now appeared to have been reduced to the more dubious single Soviet nuclear guarantee. Indian apprehensions had also risen because of the perception that the two superpowers had neutralized each other with their retaliatory-strike and assured-destruction capabilities, which tended to make potential and veiled Chinese nuclear threats appear more credible. The 1974 Indian atomic test may, therefore, be seen as the delayed Indian response to evolving global realignments. The timing of the test also appeared to confirm that the "go ahead" may have been given by Prime Minister Gandhi sometime in the spring of 1972, just after the visit of Nixon to Beijing in February of that year. Earlier statements from the Indian Department of Atomic Energy had indicated that it would take about eighteen months to move from a decision to conduct an atomic test to the actual detonation.

The timing of the test also suggests a more immediate domestic rationale. With the signing of the Simla Agreement between India and Pakistan in 1972 and simultaneous Indian efforts to normalize relations with China, the urgency of exercising the nuclear option had receded. From the external standpoint, there may have been greater disadvantages than benefits in carrying out an atomic test at this stage. On the other hand, by the spring of 1974 Mrs. Gandhi faced growing economic and political problems at home. Inflation, monsoon failures, and the oil crisis had produced a sluggish economy.[48] The detonation of an atomic device in May 1974 may have been directed at a domestic audience to dampen the stirrings of political unrest and to generate national pride and solidarity.

Whatever the underlying motives for the test, the decision accelerated a Pakistani commitment to develop what Prime Minister Bhutto called a "Muslim bomb." According to Indian assessments, the Pakistani nuclear bomb is allegedly being developed with indirect Libyan and perhaps Saudi financing, uranium supplies from Niger, and the clandestine transfer of equipment from Western Europe and North America.[49] The Pakistani-Libyan Holding Company, a joint company set up in late 1978 by the Libyan and Pakistani governments for promoting industrial developments in Pakistan, was considered by some Indian observers to be the conduit for Libyan financing of Pakistan's nuclear bomb program. The Libyans, in turn, had purchased about 450 metric tons of uranium concentrates or yellow cake from the government of President Seyni Kountche of Niger, and more purchases were considered likely. Ironically, the Libyans under Colonel Qadaffi had initially approached the Indian government in the summer of 1978 for expertise and assistance in manufacturing an "Islamic bomb." In exchange, Libya offered India considerable financial assistance. The Janata government declined the deal. Meanwhile, various related equipment for setting up a uranium enrichment plant and a spent-fuel reprocessing facility were imported by Pakistan from Britain, Switzerland, and West Germany. Finally, reports submitted by U.S. Senator Alan Cranston indicated that China was helping Pakistan develop the vital centrifuges used in the Kahuta plant's uranium enrichment process.[50] These plants when operational would provide Pakistan with separate routes to nuclear weapons capability.

The implications of these developments are far-reaching. Even if India's present dilemma stems from its own decision to test a

[48] In 1973, the consumer price index had risen by 17.8 percent over the previous year, while the food price index had risen 21.3 percent. In 1974, inflation in consumer prices was running at 28 percent and in food prices at 31 percent, largely because of the drastic increases in the price of oil imports and the failure of monsoon rains in 1973. Figures obtained from Surjit S. Bhalla, "India's Closed Economy and World Inflation," in William R. Cline and Associates, *World Inflation and the Developing Countries* (Washington, D.C.: Brookings Institution, 1981), p. 137.

[49] See *Hindustan Times*, December 21, 1980; and the *Washington Star*, April 14, 1981.

[50] See reports in the *Statesman*, June 23 and 24, 1984. See also earlier reports in the *New York Times*, April 28 and 30, 1981; and the *International Herald Tribune*, May 14, 1981.

"peaceful" nuclear device, clearly New Delhi may now find that the traditional policy of only threatening to exercise its option can no longer be sustained. Pakistan's current plans imply that both India and Pakistan are already engaged in a latent nuclear arms race.

By early 1981, there were demands in India by members of the government and the opposition, defense analysts, and the news media that the country should openly develop nuclear weapons.[51] These calls took on more urgency after the Reagan administration decided to arm Pakistan with sophisticated weapons that included the F-16 combat aircraft, which is capable of serving as a nuclear weapon delivery system against India. The American decision was part of its Persian Gulf strategy in response to the Soviet invasion of Afghanistan. Unlike the Carter administration's earlier position to provide modest arms to Pakistan but also to continue to follow a strict antiproliferation policy against it, the Reagan administration chose to provide military assistance with no explicit restrictions placed on Pakistan's nuclear activities.

These trends prompted K. Subrahmanyam, a defense specialist in India, to demand that India openly amass a nuclear weapons arsenal. According to Subrahmanyam, an Indian nuclear weapons program would not only preempt the Pakistani nuclear weapons program but would make "the United States realize that they cannot ignore a nation of 700 million people with nuclear weapons just as Nixon said about China in 1971. It will enable us to deal with China on an equal basis. We shall be able to sustain our cordial relations with the Soviet Union."[52]

Should the government of India choose to proceed along the seemingly irrevocable path of a nuclear weapons program, it would have to take into account the effect on both a nuclear China and a nuclear Pakistan, especially in the context of the quasi-alliance relationship that exists between them. The question then is whether an Indian decision to embark on a nuclear weapons program would enhance or reduce Indian security. There are arguments on both sides. In the case of China, the traditional Indian posture of threatening to exercise the nuclear option seemed to ensure that China's

[51] See the variety of views on the nuclear issue presented in a special issue of *World Focus* (New Delhi), June 1981, entitled "Must India Have the Bomb? A Debate." Contributors included K. Subrahmanyam, T. M. Kaul, P. C. Lal, Girilal Jain, and other prominent Indian Foreign Service personnel.

[52] Subrahmanyam's recommendation is reported in the *New York Times*, April 30, 1981. For a collection of Indian views on nuclear policies, see *Nuclear Myths and Realities: India's Dilemma*, ed. Subrahmanyam.

nuclear arsenal would continue to be directed at the Soviet Union instead of at India. Under the proposed new Indian nuclear policy, some of China's nuclear weapons would now surely be directed at India, against whom China would in fact possess a credible deterrent based on a retaliatory-strike capability or even a first-strike capability against India's nuclear installations. In 1984, China did not appear to possess either second- or first-strike capability against the Soviet Union. The retaliatory damage caused by the few Chinese missiles that may escape a Soviet preemptive first strike may be perceived by the Kremlin to be of an acceptable level.

The counterargument suggests that the growth of India's rocket and space programs, which will in effect provide India with Intermediate Range Ballistic Missile (IRBM) capability, would soon produce a condition of mutual deterrence between China and India.[53] Even if Indian deterrence against China does not appear credible, the situation is not fundamentally different from the prevailing condition, of India's dependence on Soviet nuclear guarantees against a China supported by the United States—a situation that unhappily makes the Soviet nuclear guarantee weak. On the other hand, a nuclear India backed by a friendly Soviet Union and a nuclear China backed by a friendly United States would produce a symmetrical relationship between India and China that would be more stable.

In the case of Pakistan, the argument against producing nuclear weapons rests on the greater likelihood of an Indo-Pakistani nuclear war regardless of rational policies of deterrence pursued by either side. Given the intensity and emotion surrounding Indo-Pakistani disputes and given the history of three wars between the two countries, nuclear weapons are more likely to be used under the conditions of mutual paranoia that exist. Even if a stable mutual deterrent situation could be established in theory (in which both India and Pakistan would possess retaliatory-strike capability), in practice deterrence could break down under conditions of extreme stress. Instead of pure deterrence, both sides might be tempted to adopt a nuclear war–fighting posture that could in fact lead to nuclear war. Moreover, these weapons could be directed at civilian targets and

[53] Rodney Jones estimated that by the year 2000, India could possibly deploy twenty to thirty IRBMS as a counter-city deterrent capable of reaching China's urban interior if not its Pacific coast. See Rodney Jones, "Small Nuclear Forces," *The Washington Papers*, no. 103, Praeger Special Studies (New York: Praeger Publishers, 1984), p. 44.

would cause large-scale human destruction and suffering if nuclear war were to occur.

The argument in favor of a nuclear weapons program in India rests on the assumption that Indian and Pakistani decision makers are no less rational and responsible than Soviet, American, or Chinese decision makers. The development of nuclear weapons by both India and Pakistan might eliminate direct wars, both nuclear and conventional, between the two states.[54] The parallel here would be the experience of the United States and Soviet Union, which have not engaged in direct conventional wars with each other because of fear of escalation to the nuclear level.

A nuclear arms race on the subcontinent would be decisively in favor of India. It would resemble the nuclear arms race between the Soviet Union and China, in which the Chinese have neither second- nor first-strike capabilities against the Soviets while the Soviets have both. As in the Soviet-Chinese nuclear relationship, the Indo-Pakistani nuclear relationship would be one in which Pakistan would not dare attack because India would carry retaliatory-strike capability, while India would not need to attack although it might have a disarming first-strike capability. The situation would be one of Indian nuclear dominance and control.

Moreover, if the Pakistanis are perceived to be developing an "Islamic bomb" for extended deterrence or for the defense of Arab Muslim states, Pakistan would also face a nuclear threat from Israel, who might be tempted to preempt before the nuclear weapons program got fully under way. The Israeli raid on Iraq's Osirak nuclear reactor in 1981 is a demonstration of possible Israeli reactions to an overt or covert nuclear weapons program that is perceived to endanger Israeli security.

On balance, it would appear that the risks of nuclear war on the subcontinent would be increased if India embarked on a nuclear weapons program. Although a state of mutual nuclear deterrence might prevail between India and China, preemptive strikes by Pakistan followed by retaliatory strikes by India would be more likely to occur between India and Pakistan. Even if India is able to maintain a clear lead in the nuclear arms race with Pakistan and even if it is able to knock out Pakistan's nuclear facilities at any time, Pakistan under such conditions might be tempted to launch an attack in an-

[54] See T. T. Poulose, *Nuclear Proliferation and the Third World* (New Delhi: ABC Publishing House, 1982), pp. 4–12; and Donald L. Clark, "Could We Be Wrong?" *Air University Review*, no. 20 (September-October 1978):28–37.

ticipation of such Indian actions. They might believe that a preemptive strike on Indian nuclear facilities would at least reduce the damage that would occur from an Indian retaliatory strike.

The tragedy of the present situation is that there appears to be no pragmatic way out of this nuclear labryinth—unless Pakistan decides to stop further moves toward acquiring a nuclear weapons capability unilaterally and India pledges not to undertake any further nuclear tests. Such a pledge by India would be difficult given the continuing growth of Chinese nuclear power. A renunciation would imply that India would give up even its traditional strategy of maintaining a nuclear weapons option. Under these circumstances, current Indian policy assumes that the China-India-Pakistan nuclear chain would be difficult to break. Thus whether the civilian nuclear energy and space programs are viable or not, the continuation of these programs remains essential from a defense standpoint.

What the traditional and revised nuclear policies indicate is that perceptions of the external nuclear threat have merely intensified from the first phase to the second. The first phase was one in which a strong Chinese nuclear threat was perceived but the ability to respond with a credible Indian nuclear deterrent policy was low or nonexistent. The second phase is one in which India perceives a two-pronged nuclear threat: a growing Chinese nuclear arsenal and IRBM/ICBM delivery capability and a covert Pakistani nuclear weapons program. The nuclear threat is therefore perceived to be even greater in the present circumstances. However, India has the technological capability and potential in its nuclear and space programs to respond to both threats with a credible nuclear weapons policy of its own. The constraint on such a policy is the prospective international countermeasures that may be adopted by the major powers who wish to prevent horizontal nuclear proliferation.

CHAPTER THREE

Internal Security and the Use of Force

ALTHOUGH India has always faced serious problems of internal strife and armed insurgency, these issues began to receive greater attention after the National Emergency was declared by Prime Minister Gandhi in June 1975. Security as understood in India today is the absence of threats to the territorial integrity and the political and economic stability of the Indian Union. In the internal context, secessionist movements through acts of terrorism or guerrilla warfare, civilian communal violence that mainly involves religious but also linguistic and caste conflict, and other violence arising from economic and political issues are all considered threats to the state from within. This broadly based concept is reflected in the mandate and composition of the Political Affairs Committee of the Cabinet. The committee, established in the mid-1970s, includes almost every cabinet minister in government: Defence, Home Affairs, External Affairs, Finance, Industry and Labor, Railways, Commerce, and Steel and Mines.

Problems of internal and external security converge when the nation's armed forces are trained and geared both to defend against external aggression and to maintain internal law and order.[1] Most democratic states, however, prefer to use two separate bodies of armed forces for the two forms of security. This is also true in India, where internal security is for the most part maintained by the local police, and at times by specially trained armed constabulary and paramilitary forces. Responsibility for internal security is also assigned to a separate ministry in India, the Ministry of Home Affairs. Nevertheless, as internal violence escalates, the probability of intervention by the armed forces under the Ministry of Defence also increases. This trend was evident in the early 1980s when internal strife broke out in Assam, Gujarat, Maharashtra, and Punjab and

[1] For a discussion on the interaction between the two levels of security, see P. R. Chari's "Internal Dimensions of National Security," *Strategic Analysis* (New Delhi) 3, no. 7 (October 1979):249–53; and his "Internal and External Security," *Economic Times* (Bombay), October 26, 1980.

the army was called in to deal with the civilian violence. Meanwhile, the army continues to be deployed against armed insurgencies in Nagaland, Mizoram, Meghalaya, Manipur, and Tripura, where there are sustained or sporadic efforts to achieve independent statehood. Whether it involves the use of the police or the military, the use of force to curb internal political unrest and violence tends to generate far greater domestic controversy than does the use of the armed services to defend the country against external aggression.

In this chapter, the discussion will deal first with the nature and the causes of internal political violence, and then will consider the instruments and applications of force to restrain or eliminate such violence. The political process for security policy making is the subject of chapter four.

I. Issues in Internal Security

The Nature of the Problem

The causes of domestic violence are complex and varied, as are the methods adopted by dissident factions to achieve their many territorial, political, and economic objectives. There are four basic forms of internal violence that many Third World nations have had to cope with: (a) civil war or the threat of it; (b) guerrilla warfare and other forms of armed insurgency; (c) terrorism; and (d) civil strife arising from ethnic antagonisms, industrial strikes, or from political dissent on various contemporary issues and policies.

Civil war prevails when armed camps of opposing factions within the same country are engaged in battle to seek the control of the state's government or some territory within the state or to achieve some other political, economic or military objective. The belligerent forces on either side usually have some military training and are commanded by a hierarchy of officers who plan strategies to achieve desired military and political objectives.[2] The conduct of war here is not unlike wars between independent nations. Civil wars may lead to the secession of a piece of territory constituting a new independent state or to the maintenance of the old state under the government of the victorious faction.

[2] According to David Wilkinson, a revolutionary civil war is a "war between belligerent factions seeking by organized violence to acquire a monopoly of force in a state, to establish their rule over it, and to conquer their opponents" (Wilkinson, *Revolutionary Civil War: The Elements of Victory and Defeat* [Palo Alto, Calif.: Page-Ficklin Publications, 1975], p. 1). This book is an excellent study of twentieth-century civil wars in China, Russia, Spain, and Mexico.

By this definition of civil war and in comparison with the civil wars of Mexico, Russia, Spain, and China, there has been no major civil war in India in the twentieth century, either before or after Independence. But conditions that bordered on it prevailed when independence was achieved in 1947. Muslim aspirations for separate statehood led to widespread Hindu-Muslim bloodshed during the Indian nationalist movement in the 1930s and 1940s. Indeed, the partition of India and the creation of a new state of Pakistan as a homeland for Indian Muslims was intended to ward off what appeared to be an impending civil war between the Hindu and Muslim communities. It may be argued that civil war between the Hindu and Muslim communities—represented by the Indian National Congress, led by Jawaharlal Nehru, Mohandas Gandhi, and Vallabhai Patel, and by the All-India Muslim League, led by Mohammed Ali Jinnah—was avoided only by definition. War between India and Pakistan over Kashmir in December-January 1947–1948 may be perceived as a civil war that was raised to the level of an international conflict by the recent partition of the subcontinent into two new independent states. The conflict involved the armed forces of both India and Pakistan that until 1947 had constituted the united armed forces of British India. The 1947–1948 Indo-Pakistani war was not fundamentally different from the 1950–1951 Korean War and the prolonged war in Vietnam, in which the countries involved had also already been partitioned.

Civil war took place in Pakistan in 1971 between West Pakistan and East Pakistan, but here too the situation suggests some ambiguity. Through much of 1971 after the breakdown of negotiations in March between Sheikh Mujibur Rahman's Awami League party and the military regime in West Pakistan, sporadic guerrilla warfare was carried on by largely untrained Bengali resistance groups against the Pakistan army. When war broke out in December of that year between India and Pakistan, there were concurrent elements of civil war, since the now armed and well-trained Mukti Bahini, the main Bengali resistance group, coordinated their military operations with the Indian army to defeat the Pakistan army in East Pakistan. However, civil war in India involving regular armed forces on either side has not occurred since Independence.

Guerrilla warfare occurs when armed groups conduct sporadic or sustained attacks on targets of military significance in order to undermine the control of the existing authority or government in power to achieve some territorial, political, economic, or social objective. The difference between civil war and guerrilla warfare lies in the scale and the nature of the operations. Civil war suggests confrontations between opposing conventional forces on a much larger scale than guerrilla war. Guerrilla warfare suggests sporadic or sustained hit-and-run tactics by armed insurgents. The military capabilities of the guerrillas are much lower than those of the established authority, but guerrilla tactics make it difficult to deploy conventional forces against them on a substantial or sustained scale. Guerrilla warfare may be a part of the civil war, as was the case in China and Vietnam.

Guerrilla warfare has taken place over extended periods in the northeast sector of India in Nagaland, Mizoram, Meghalaya, Manipur, and Tripura. Occasional and brief Communist-inspired insurgencies, known as the "Naxalite" movement, have occurred in West Bengal, Andhra Pradesh, and Kerala. Counterinsurgency operations have been conducted by the Indian army against this form of violence and the extent of their success or failure will be considered below.

The violent methods of terrorism employ the weapons of fear and publicity. Unlike guerrilla warfare, which is primarily directed at military or strategic targets, terrorism is aimed at civilians, often indiscriminately. The targets—school children, airline passengers, notable businessmen, or politicians—may be irrelevant to the ultimate objectives of the terrorists, but attacking them produces fear and extensive publicity, which is expected to highlight the cause or demands of the terrorists among the public. A small group of people is thus able to command far more power and attention than their numbers would indicate. By the combination of fear and publicity they hope to force governments or others in authority to capitulate to their demands. Although terrorist activity is not new, it has seized the world's attention in recent years.[3]

Extensive references to terrorist activity in India began in 1982 with the renewal of Sikh demands for greater political and economic autonomy. The more extreme elements have gone further and demanded an independent Khalistan. The targets of Sikh ter-

[3] See Connor Cruise O'Brien, "Liberty and Terrorism," *International Security* 2, no. 2 (Fall 1977):56–67. See also David Fromkin, "The Strategy of Terrorism," *Foreign Affairs* 53, no. 4 (July 1975):683–98; Paul Johnson, "The Seven Deadly Sins of Terrorism," and Walter Lacquer, "The Futility of Terrorism," in *World Politics Debate: A Reader in Contemporary Issues*, ed. Herbert M. Levine (New York: McGraw-Hill, 1983), pp. 248–57; Yonah Alexander, David Carlton, and Paul Wilkinson, eds., *Terrorism: Theory and Practice* (Boulder, Colo.: Westview Press, 1979); Richard Clutterback, *Guerrillas and Terrorists* (London: Faber and Faber, 1977); Paul Wilkinson, *Terrorism and the Liberal State* (New York: John Wiley, 1977); and Walter Lacquer, *Terrorism* (Boston: Little, Brown, 1977).

rorists have been mainly Hindu politicians associated with Mrs. Gandhi's Congress party, including the leader herself—whether in fact they are the cause of the Sikhs' problems or not—as well as other Sikh moderates who oppose their demands and strategy. Their targets have also included bus passengers and other innocent civilians whose deaths have generated both fear and publicity. The hijacking of government-owned India Airlines passenger planes has also proved to be an effective strategy for both Sikh and Muslim Kashmiri separatists in India at least to gain national and international publicity for their causes.

The strategy of terrorism is more likely to succeed in democratic societies where the principles of liberty restrain governments from resorting to authoritarian means to curb terrorist activity. Given the arduous and complex—even chaotic—nature of the democratic process in India, official efforts to curb terrorism have tended to result in an erosion of the democratic process. Terrorism seems to call for bypassing constitutional and judicial processes and resorting to special ad hoc measures. This tendency has become apparent in government efforts to restrain terrorism by Sikh extremists in the Indian Punjab. Terrorist activity there during 1982–1984 has also allegedly been aided by the ability of the terrorists to obtain arms and sanctuary from the neighboring Punjab province of Pakistan. A similar situation may be seen in neighboring Sri Lanka where Tamil separatists indulging in acts of terrorism against the majority Sinhalese population were able to obtain arms and sanctuary from their co-ethnic sympathizers in the Indian state of Tamil Nadu. To the extent that terrorist groups are able to obtain external political or military support, the situation becomes one of surrogate warfare that could deteriorate into international confrontation and lead to conventional wars.[4]

The success of terrorism depends not only on the generation of fear and publicity but on the reaction of the authorities. Especially in the case of territorial secessionists, government reactions that attempt, for instance, to round up at will suspects belonging to the ethnic group tend to further alienate all or most members of that group. For example, the success of the terrorist activity by the Sikhs in the Punjab who seek an independent Khalistan has depended on the reaction of the Indian government. Sikh extremists expected that police retaliation against Sikh suspects or the resort to armed force to destroy terrorist strongholds would win over Sikh moderates to their cause. This objective was partially realized when the

[4] See Lacquer, "The Futility of Terrorism," p. 256.

government ordered the army to attack the Golden Temple in Amritsar, where armed extremists led by Jarnail Singh Bhindranwale had taken refuge, in June 1984. The temple had become the operations center of the Sikh extremists for the conduct of their acts of terrorism, which had alienated them not only from the rest of the Indian population but also from the majority of the Sikhs as well. The Indian military action, however, produced worldwide outrage among Sikhs, who felt that their holiest place of worship had been defiled by the Indian government. The government retaliation provided by the Indian army attack and the "martyrdom" of Bhindranwale did more to advance the cause of the Sikh extremists than their earlier acts of terrorism.

Thus, in democratic systems, dealing with terrorist acts poses a dilemma. Terrorist tactics seem to call for resolute and harsh measures by the government in order to inform all existing and potential terrorists that their strategy will not pay. On the other hand, if such harsh measures mean the arbitrary arrests of many innocents and severe police or military retaliation that bypasses normal constitutional and judicial procedures, the danger exists that the terrorism stemmed by these measures will be replaced by government-conducted terrorism against its own people.

The nature of civil strife is varied and is probably the most persistent and widespread cause of violence in India. Such civilian violence is rooted in religious, linguistic, caste, and tribal differences, and in political, economic, and social rivalries. Hindu-Muslim communal rioting continues at various levels of magnitude; in some years such rioting may cause over a thousand deaths. Other violent conflict has occurred between Telugu- and Tamil-speaking peoples in Andhra Pradesh; Assamese and Bengalis in Assam; Maharashtrians and South Indian settlers in Bombay; and caste Hindus and Harijans (the former "Untouchables") in various parts of India, especially Gujarat and Bihar. Industrial unrest has also led to violence between workers and police. Labor strikes, especially those that occur on a national scale as in the case of railway strikes, can paralyze the economy. Political dissent has often produced calls for a "hartal" or a "bandh," that is, the cessation of all commercial and administrative activity in a city as a protest against certain policies of the government. Such calls have invariably led to violence between those who favor and those who oppose the moves, as well as between demonstrators and police.

All of the above conditions—the threat of civil war, guerrilla warfare, terrorism, and violent civil disturbances—are considered in

India to be a threat to the internal security of the state. In the case of guerrilla warfare and terrorist activity, the "enemy" is armed; in the case of violent political demonstrations, the "enemy" is often unarmed or inadequately armed. Indian policy makers are aware that such conditions could lead to the disintegration of the state; two major experiences on the subcontinent bear witness to the possibility. The partition of India and the creation of Pakistan in 1947 was the result of civil strife between Hindus and Muslims during the independence movement and the seemingly irreconcilable political differences between the Indian National Congress and the All-India Muslim League. Similarly, the separation of Bangladesh from Pakistan in 1971 was the result of the internal revolution, civil war, and guerrilla warfare in East Pakistan, eventually aided by Indian military intervention.

Problems of internal security are also linked to India's problems of external security. Hindu-Muslim riots and the separatist movement in Kashmir are related to issues that date back to the pre-Independence "Pakistan" movement and the subsequent partition of India in 1947. Continuing occurrences of such communal rioting in India have produced Pakistani criticisms, Indian accusations of Pakistani interference in its domestic affairs, and a general deterioration in Indo-Pakistani relations. Insurgencies and separatist movements in northeast India among the Nagas and the Mizos, for example, have been aided by China and Pakistan (by the latter mainly before the creation of Bangladesh). Political disturbances in Kashmir constantly face the prospect of overt or covert Pakistani intervention. As noted earlier, Sikh terrorist activity for an independent Khalistan allegedly received some Pakistani matériel support, although Pakistan has denied such allegations. The "Naxalite" Communist movement of the 1960s, which involved acts of terrorism and sabotage, drew its inspiration from the principles of Maoist guerrilla warfare and its weapons from Maoist China. The Dravidistan separatist movement in Tamil Nadu—the creation of an independent nation comprising the Dravidian states of Tamil Nadu, Andhra Pradesh, Karnataka, and Kerala—did not involve much violence. But separatist sentiment remains an important factor that must be taken into account in the central government's national integration policies.

Problems of Internal Security

In general, problems of internal security in India may be divided into two broad categories: secessionist and nonsecessionist. The first type includes acts of violence arising from demands in varying degrees for territorial secession from India. The second type includes violence that stems from internal political issues but has no planned separatist objectives. Both constitute threats to the integrity of Indian Union. Secessionist movements suggest the eventual "Balkanization" of India, one of the major dangers faced by this extremely heterogeneous republic. Nonsecessionist violence suggests dangers of internal political, economic, and social disintegration that would make the normal conduct of government difficult or impossible.

Some examples in the first category are the Kashmir issue, the Khalistan movement, the Dravidistan movement, and independence movements in Nagaland, Mizoram, Meghalaya, Manipur, and Tripura. Secessionist movements invariably depend on external overt or covert military support to sustain their armed insurgency or acts of terrorism and therefore they tend to be confined to the border areas. Examples in the second category are the early linguistic rioting in the 1950s for the separation and creation of the states of Andhra Pradesh and Maharashtra from the greater Madras and Bombay presidencies, which then included the present states of Tamil Nadu and Gujarat; the Punjabi Suba issue of the 1960s, which called for the separation of Hindu majority areas from Punjab into a new state called Haryana; and the more recent violence in Assam arising from the immigration of Bengalis into that state from Bangladesh and West Bengal. Communal conflict and violent industrial strikes may also be included in the second category. Acts of violence under this category tend to be more spontaneous and to lack organized, sustained, and armed paramilitary-type operations.

The most familiar issue in the first category is the problem of the state of Kashmir. This state, which has a Muslim majority and whose Hindu ruler acceded to India at the time of partition, has experienced periodic political turmoil initiated mainly by the Plebiscite Front and other Muslim organizations such as the Jamiat-e-Islami and its militant youth wing, the Jamiat-e-Tulaban. Domestic political disturbances in Kashmir have received the political and sometimes the clandestine armed support of Pakistan. Two wars have been fought over Kashmir between India and Pakistan in 1947–1948 and 1965 and on both occasions the problem of controlling internal conflict also had to be managed.

The persistence of the dual internal-external security problem of Kashmir may be seen in the call by the Jamiat-e-Islami in August 1980 for an "Iran-type revolution for the liberation of Kashmir from illegal occupation and enslavement by India."[5] The president of the Jamiat-e-Tulaban, Sheik Tajamul Islam, declared that Kashmiri Muslims have never accepted the "the imaginary theory that Kashmir is an integral part of India." He further added that if a plebiscite were held, the majority of the Kashmiris would give allegiance to Pakistan and not to India. In July 1984, the fall of Farooq Abdullah's National Conference government in Kashmir again led to fears of internal civil strife and countermeasures by the central government.[6] According to government claims, the state government of Farooq Abdullah had failed to curb the antinationalist activities of the pro-Pakistan Jamiat-e-Islami and Jamiat-e-Tulaban and had allegedly given assistance to Sikh extremists in neighboring Punjab. Such threats stemming from within and outside Kashmir have resulted in the presence of both a large standing police force together with paramilitary forces and an even larger deployment of army divisions. Although the population of Jammu and Kashmir is only 5 million (of which 3.5 million are Muslims), about 45,000 paramilitary forces and about 100,000 soldiers are stationed in the state to maintain law and order and to stop any threatened invasion by Pakistan.

In the Punjab, the movement for a separate state of Sikhistan had been voiced by the Akali Dal party at the time of the Pakistan movement in British India. Subsequently, the demand has been for a "Punjabi Suba," a state to be set up within India along linguistic lines. Such a reorganization of the Indian Punjab based on majority Punjabi-speaking areas was expected to leave the Sikhs with a religious majority within the proposed new boundaries. These demands were met in 1966 when those areas where Punjabis claimed Hindi as their language were separated into the state of Haryana, leaving a Sikh majority of roughly 55 percent in the now further divided Punjab. The demands were virtually conceded to as a reward for the valor and sufferings of the Sikhs during the 1965 Indo-Pakistani war.

With most of the Akali Dal demands having been met, much of the pressure for an independent Sikh nation appeared to have been eliminated until a series of newer and more extreme Sikh demands began in the early 1980s. The moderates under the Akali leader Harchand Singh Longowal desired greater political and economic autonomy for the Punjab; the extremists under Bhindranwale, sup-

[5] See *Foreign Broadcast Information Service* 8, no. 154 (August 7, 1980):E3–E4.

[6] See *New York Times*, July 5, 1984; and *Statesman*, July 3, 5, and 9, 1984.

porting the same demands, resorted to widespread acts of terrorism to achieve these ends. Bhindranwale claimed never to have endorsed an independent Khalistan state for the Sikhs but also claimed that he did not oppose one. Certain factions within this group, however, including the All-India Sikh Students Federation, Dal Khalsa, and the Dashmesh Regiment, openly claimed that their objective was to create an independent state. The pro-Khalistan factions in India were supported by wealthy Sikhs in Britain, Canada, and the United States. Most prominent among the overseas Sikh leaders was a physician based in Britain, Jagjit Singh Chauhan, who through his National Council of Khalistan provided the inspiration and ideology for Bhindranwale's extremist movement.[7] Other overseas groups included the Dal Khalsa in Britain and West Germany and the Babbar Khalsa and Akhand Kirtani Jatha in Britain and Canada.

Between August 1982 and May 1984 almost 600 people were killed by the terrorists, 200 alone in the two weeks before June 6, 1984.[8] The victims included Hindu as well as moderate Sikh politicians, businessmen, and other civilians. There was also sporadic killing of policemen, both Hindu and Sikh, who were perceived as symbols of government authority. On June 5 Indira Gandhi's government ordered Indian army units commanded by a Sikh, Lt. Gen. Ranjit Singh Dayal, to attack the fifteenth-century Golden Temple, the holiest of the Sikh shrines, from which Bhindranwale's group launched its terrorist activities. Initial government reports indicated that Indian army operations against the temple caused the deaths of 576 people, including 84 soldiers. Subsequent government reports indicated the deaths of 492 Sikh militants and 93 soldiers, including 9 officers. However, unofficial reports put the number of deaths at over a thousand, including over 100 soldiers and several innocent Sikh pilgrims caught in the cross fire.

The long-term consequences of the Indian military attack remain yet to be seen. Although only a minority of Sikhs supported the Bhindranwale tactic of violence, the perceived desecration of the shrine on June 5 suddenly alienated the moderates under Longowal as well as several prominent Sikh community leaders who had been in favor with Mrs. Gandhi. If this alienation continues and more Sikhs join the Khalistan movement, then the tactics of the Bhindranwale group will have succeeded and their objectives will have

[7] See *New York Times*, June 14 and July 11, 1984.

[8] See *New York Times*, June 6, 7, and July 29, 1984.

been partly achieved. Indeed, the assassination of Indira Gandhi a few months later by two of her trusted Sikh bodyguards was a demonstration of the depth of Sikh antagonism toward her. On the other hand, Mrs. Gandhi argued that the desecration of the temple occurred when Sikh terrorists used it as a sanctuary to conduct acts of terrorism and that the army attack was ordered only after all other measures had failed.[9] The government's attitude under the late prime minister was that terrorist activities call for the suspension of basic democratic principles and the resort to limited and temporary authoritarian measures that include curfew, mass arrests and detention without trial, and restrictions on the freedom of the press, all of which are now legally permissible under the provisions of the 1980 National Security Act. These measures apparently did not reduce Sikh efforts to produce an independent Khalistan state through armed violence.

The tactics of Rajiv Gandhi's Congress government in dealing with the Punjab issue have been more conciliatory. Recognizing the failure of the hard-line approach of his predecessor, Rajiv Gandhi sought accommodation with the Akali Dal. After acceding to several immediate Sikh demands such as the release of many Sikh political prisoners and amnesty for Sikh deserters from the army, the Indian government signed an agreement in September 1985 with Longowal's Akali Dal that was intended to normalize conditions in the Punjab. As part of the process of normalization, elections were to be held in the Punjab within a few weeks. Longowal was assassinated by Sikh extremists a few days after the agreement was signed, but elections were still held. For the first time the Akali Dal, led by Longowal's successor, Surjit Singh Barnala, won the majority of the seats in the Punjab Assembly, enabling the party to form a government on its own. This result was probably desirable from the standpoint of the central government in New Delhi, even though it meant the defeat of the Congress party in the state, because it put Sikh moderates in power and undercut the influence of the extremists in the Punjab.

Other regions where separatist movements have involved insurgency, occasional acts of terrorism on civilians, and countermilitary actions are all found in the volatile northeast of India. These territories are Nagaland, Mizoram, Meghalaya, Manipur, and Tripura.

[9] For Mrs. Gandhi's justification for the army attack on the Golden Temple, see report in the *Indian Express*, June 20, 1984. For other Indian assessments, see *Statesman*, June 20, 1984; and *Hindu*, June 25, 1984.

However, unlike Kashmir, there has been no external threat of overt, direct military intervention by China or Pakistan; external interventions here have been primarily subversive through the supply of arms, matériel, and training.

The first of such major problems in the northeast arose from the demand for independence by the Naga tribes, led by Gen. A. N. Phizo. The Naga demand for a separate state dates from the years immediately after Independence and has been supported by western missionaries who had converted much of the Naga population to Christianity. Nagaland was expected to comprise all of the Naga-inhabited areas in the Naga hills, which in 1981 included 780,000 people.[10] The insurrection was helped by its geographical location; the area borders the Kachin district of North Burma, where a similar insurrection was conducted by the Kachin Independent army against the Burmese army. The Kachin district in turn borders China's Yunan province, and the Kachin insurgents have consequently been able to provide the Naga insurgents with access to Chinese arms and provisions. According to Lt. Gen. A. M. Vohra, "the Chinese assistance by way of arms and ammunition, as well as training camps has kept the insurgency going in Nagaland for years."[11]

The objective of the Naga rebels to establish an independent Nagaland through armed insurrection first brought the Indian army to this region in 1956. Much of the rebellion subsided in 1961 when the government announced its decision to grant the area separate statehood within the Indian Union. In September 1964, the "Suspension of Operations Agreement" came into force, but even after this some have continued to struggle for independence. In 1975, the government and the Naga rebels signed the Shillong Agreement for a cease-fire. Nagaland has remained largely peaceful since, except for a few incidents of raids by pockets of Naga underground resistors from their bases across the Burma border.[12] The Indian army continues to be deployed in Nagaland to deal with these sporadic attacks.

A similar situation existed among the Mizo tribals, who number

[10] Population figures are from table 1.2 in *India: 1981* (New Delhi: Publications Division, Ministry of Information and Broadcasting, 1981), p. 8.

[11] See Lt. Gen. A. M. Vohra, "The Anatomy of Insurgency," *Tribune*, May 23, 1980. See also Arul B. Louis, "Insurgency in the Northeast," *Hindustan Times*, January 7, 1981.

[12] See *Report: 1982–83* (New Delhi: Ministry of Home Affairs, Government of India, 1983), p. 8.

some 500,000. Most of them, like the Nagas, have been converted to Christianity by western missionaries. The Mizo hills have similar geographical advantages for conducting guerrilla warfare, and the region has common borders with Burma and Bangladesh (earlier East Pakistan). Before the creation of Bangladesh, Pakistan had assisted the movement with training and supplies. The Mizo demand for independence began in the early 1960s and has been led by the Mizo National Front (MNF). Military operations against the Mizo insurrection began in March 1966 and continued until June 1976, when the territory was separated from the state of Assam and named Mizoram with the status of Union Territory.[13] In 1982, the MNF renewed its struggle for an independent Mizo state. The group was then banned and the central government ordered a major military crackdown on the guerrillas.[14]

Like Nagaland, the Mizo issue has not been completely resolved. The leader of the Mizo rebels, Laldenga, has continued to hold out the threat of insurgency unless two basic problems are resolved. First, the Mizos feel that the status of Union Territory is inadequate and have demanded full statehood and a more autonomous relationship between Mizoram and New Delhi. Second, they want various steps taken for the "psychological rehabilitation" of the underground rebels and guarantees made for the "survival of the Mizos as a distinct ethnic and cultural entity."[15] In November 1985 a tentative agreement was reached between the Indian government and Laldenga on these demands of the MNF.

Three other regions in the northeast—Meghalaya, Manipur, and Tripura—have had varying degrees of insurgency. In what is now Meghalaya, a state with a population of 1.3 million, guerrilla warfare conducted by Khasi and Garo tribesmen against government troops ended after this state was created in the 1960s out of Assam along the Bangladesh border. About the same time, the states of Tripura on the Bangladesh border and Manipur on the Burma border, with small populations of 1.5 million and 2 million respectively,

[13] Such territories do not have full statehood and are under the ultimate jurisdiction of the central government, though they possess considerable local autonomy. Other examples of Union Territories in India are the national capital of Delhi and the former Portuguese and French territories of Goa and Pondicherry.

[14] See *Times of India*, January 23 and 24, 1982.

[15] *Times of India*, February 13, 1980. See also the statement by the Mizoram representative in the Lok Sabha, R. Rothuana, during the debate on the budgetary allocations to the Ministry of Defence (*Lok Sabha Debates*, 6th ser., vol. 12, no. 24 [March 27, 1981], cols. 417–19).

were also set up to mitigate separatist tendencies in these regions. But the insurgencies and counterinsurgency measures continued, taking on significant dimensions in Manipur and Tripura in 1976 and 1980; Nagaland, Mizoram, and Meghalaya had by then already experienced decades of bloodshed.[16] In Manipur, the insurgency was led by the People's Liberation Army and members of the Revolutionary Government of Manipur, which according to Indian reports had been trained almost a decade earlier in what was then East Pakistan and in China.[17] The insurgencies in Manipur and Tripura, which were on a much smaller scale than those in Nagaland or Mizoram, principally involved sporadic ambushes of Indian army personnel, political assassinations, bank robberies, and the extortion of food and money from villagers.

Separatist tendencies of less intensity have occurred in Tamil Nadu.[18] As in the case of the demands for Khalistan, an independent state of Dravidistan had been demanded by the Tamil-speaking people in what is now the Indian state of Tamil Nadu (formerly the state of Madras). Initially, the concept of Dravidistan included all four of the states in the south where Dravidian languages are dominant—Andhra Pradesh (Telugu), Karnataka (Kanada), and Kerala (Malayalam), in addition to Tamil Nadu (Tamil). But the Dravidistan movement led by the Dravida Munnetra Kazagham (DMK; or the Dravidian Progressive Federation) found few adherents outside the Tamil-speaking peoples. The DMK secessionist movement of the 1950s has since become muted; its objectives have been reduced to demands for greater autonomy and the more pragmatic goals of internal economic and social reform. The party itself split in 1972 over the issue of whether to emphasize greater internal reform or greater provincial autonomy. The new and now dominant party, the All-India Anna DMK (named after the founder of the party, Annadurai), led by South Indian film star, M. G. Ramachandran, challenged and defeated the parent DMK led by M. Karunnanidhi, who had begun to voice new demands for regional autonomy. Indeed, Karunnanidhi had even begun to project himself as the "Mujib of Tamil Nadu," a reference to Sheikh Mujibur Rahman, the Muslim Bengali nationalist who led the movement for an independent Bangladesh from Pakistan in 1971.

[16] See *New York Times*, March 13, 1984.

[17] *Overseas Hindustan Times*, June 10, 1982.

[18] For a brief discussion of these problems, see Robert L. Hardgrave, *India: Government and Politics in a Developing Nation* (New York: Harcourt Brace Jovanovich, 1980), pp. 91–95.

One other separatist movement, based on ideology rather than ethnic nationalism, should be mentioned. A Communist movement, calling for armed liberation from the Indian Union, was conducted in the West Bengal province of Naxalbari. In 1967, peasants in this strategic district seized control of much of the land from the landlords and were backed in this endeavor by Radio Beijing, which referred to the area as a liberated base from which to conduct protracted revolutionary warfare for the liberation of other areas.[19] The rebellion was led by extremists, including Charu Mazumdar and Kanu Sanyal, and the methods of the party were based on the early Maoist tenet that "power grows out of the barrel of a gun." The rebellion was put down quickly through military intervention, but the strategy of the rebels was adopted by other pro-Mao Communists in Andhra Pradesh and to a lesser extent in Kerala. The group, organized as the Communist Party of India (Marxist-Leninist), was immediately banned by the government of India as illegal because of its refusal to accept the constitutional process. The "Naxalites," as they came to be called, nearly gained a greater victory in 1968 in the Andhra hills of the Srikakulam district, when they seized land from its owners through guerrilla raids. This effort, which was again hailed by Beijing, was quickly suppressed by 1969.

The Naxalite movement had subsided considerably by the 1970s, although there continued to be limited numbers of diehard Naxalites in Andhra Pradesh, West Bengal, and Kerala. The movement was broken both by the new international respectability and peaceful posture of Mao Zedung and his successors in China and by the electoral successes of the Communist Party of India-Marxist (CPM) in Kerala, West Bengal, and Tripura. The government has responded to the remaining pockets of violence with selective armed police action.

Apart from such claims for separate independent states, there have been lesser demands that have not included secession. These include the violent riots in the 1950s for the creation of the states of Andhra Pradesh and Maharashtra within the Indian Union, and the call in the early 1980s in Assam to expel all non-Assamese from the state in order to protect the political and ethnic identity of the Assamese. As noted earlier, the first cases arose from the initial decision of the States Reorganization Commission to maintain the

[19] See Hardgrave, *India: Government and Politics*, pp. 129–30. See also the *Hindustan Times*, Dec. 30, 1980; *Tribune*, December 13, 1979; and the discussion regarding the possible revival of the Naxalite movement by N. C. Menon, "Return to Naxalbari?" *Overseas Hindustan Times*, December 3, 1981.

Tamil- and Telegu-speaking peoples in the old Madras presidency demarcated by the British and similarly the Marathi- and Gujerati-speaking populations within the old Bombay presidency. Severe rioting, which briefly saw Indian army tanks and armored vehicles rolling through the city streets of Hyderabad and Bombay, eventually brought about separate states for the Telegu- and Marathi-speaking populations.

Lingering aspects of these two problems remain. The Shiv Sena (the army of Shivaji), founded in 1966, continued to uphold the banner of Marathi-speaking nationalism in western India.[20] The movement is inspired by Shivaji Bhonsle, the founder of the Maratha empire in the seventeenth century in western and central India. Since the empire was carved out through relentless and savage battles against the declining Muslim Mughal empire and against invading Muslim Afghans, the present party's ideology carries a strongly anti-Muslim sentiment. Hindu-Muslim communal rioting in the state of Maharashtra is often provoked or aggravated by the Shiv Sena. In the case of Andhra Pradesh, the separation of this state from the greater Madras state (now Tamil Nadu) did not resolve the problem of its principal group, which now demands the further bifurcation of the state to create a new Telengana state for the Telegu-speaking populations that would split off the coastal Andhra populations. The demands of both communities continue to cause sporadic rioting and loss of life. During the Hindu-Muslim rioting in 1984 in north Bombay sparked by the Shiv Sena, the Indian army had to be deployed to keep peace between the two communities.

The Assam agitation arose not from a demand for an Assam state, which they already have, but from a fear that the immigration of Bengalis and other groups from various parts of India would begin to make the Assamese a minority in their own state. This problem has greater international political overtones than did the earlier Andhra Pradesh or Maharashtra issues, because the Assamese protests are directed at the migration and settlement of Bengalis from both West Bengal and the independent state of Bangladesh. There have been four waves of such Bengali immigration into Assam. After the partition of India, there were Bengali Hindu migrations into Assam from what then became East Pakistan. In the 1950s and 1960s, the neglect of East Pakistan by the government, located in West Pakistan, led to a large-scale influx of Bengali Muslims. In the mid-1960s, there was another wave of immigration, this time by middle-

[20] See *New York Times*, May 19–23, 1984.

class Hindu Bengalis from West Bengal who came to Assam in search of high paying, white-collar jobs in the oil industry and by Hindus from East Pakistan fleeing the 1965 Indo-Pakistani war. During the East Pakistani civil war and the struggle for the creation of Bangladesh in 1971, large numbers of Hindu and Muslim Bengali refugees also fled to Assam to escape the massacres of the West Pakistani military. As a consequence, of the 19 million people in Assam in 1980, only 10 million spoke Assamese.[21]

The continuing Assam agitations since 1979 have produced the disruption of oil production and the closure of oil refineries at Digboi, Gauhati, Barauni, and Bongaigon. The Assamese have demanded OPEC prices for Assam oil to be paid into the state government treasury and have called for the expulsion of all immigrants who entered after 1947.[22] State government failure to resolve this issue led to the declaration of President's Rule in December 1979 and to the deployment of Indian army units where there has been extreme violence.[23] After an extended period of President's Rule, constitutional provisions required the conduct of elections for the state assembly by March 1983. The main groups calling for the expulsion of Bengalis from Assam, including the All-Assam Students' Union and the All-Assam Gana Sangram Parishad, threatened to resort to widespread and extreme violence if the elections were held before the immigrant issue was resolved to their satisfaction. The elections were held in February 1984; six thousand men, women, and children died when Hindu Assamese attacked Muslim Bengalis suspected to be illegal immigrants from Bangladesh.[24] Indian army units were called in to quell the violence and to prevent further bloodshed.

As in the case of the Punjab crisis, Rajiv Gandhi sought to placate the grievances of the Assamese population, which is fearful of becoming a minority in its own state. Elections were called in Assam in December 1985, two months after the successful and largely peaceful elections in the Punjab. Unlike the tactics in the Punjab, where it

[21] *New York Times*, April 2 and 27, 1980.

[22] For various reports on these issues, see *Hindustan Times*, Dec. 30, 1980; January 10 and 13, 1981. For official assessments of the situation in Assam, see *Report: 1979–80* (New Delhi: Ministry of Home Affairs, Government of India, 1980), pp. 13–14; and *Report: 1982–83*, Ministry of Home Affairs, pp. 6–7.

[23] *Times of India*, December 11, 1979; and January 20, 1980. President's Rule involves the suspension of the state government through a presidential proclamation and direct rule by the central government until a new state government can be installed.

[24] *New York Times*, July 25, 1984.

hoped the Akali Dal would win, in Assam the Congress party worked hard to win the elections. Its principal opponent was the Asom Gana Parishad (the Assamese Peoples' party), a coalition of various factions demanding the expulsion of nearly all Bengalis from the state, especially Muslims from the neighboring country of Bangladesh. For the first time the Congress party lost the Assam state elections. The victory of the Asom Gana Parishad raised some fears that there would be more violence, particularly if the state government decided to carry out its threat to expel most Bengalis from the state.

Probably the most important and internationally significant cause of violence in India has been religion rather than language. This has primarily involved conflict between Hindus and Muslims. To a certain extent, insurgencies in Nagaland, Mizoram, and Meghalaya can also be considered religious because most of their members are Christians. However, these regions have not been involved in interreligious bloodshed and religion is only a partial basis for insurgent demands. Moreover, no external power has supported these movements for religious reasons, although some Christian missionary groups have been supporters or sympathizers. On the other hand, tension between Hindus and Muslims has major significance in India and abroad.

The Muslims of India, numbering 85 million people in 1983, constitute the third largest Muslim population in the world after Indonesia and Bangladesh. Since the breakup of Pakistan in 1971, the Muslim population of Pakistan is slightly smaller and that of Bangladesh slightly larger than that of India. Hindu-Muslim riots in India have continued to produce Pakistani protests and Indian objections to interference in its domestic affairs. Although armed force may be used to subdue and resolve the Naga and Mizo insurrections, the nature of Hindu-Muslim violence is such that it cannot be resolved through the application of police or military forces. The Muslim population is spread throughout India, and unlike the Naga and Mizo cases, Hindu-Muslim violence takes no organized and systematic form. It tends to be spontaneous, sporadic, and scattered, which makes police control difficult and prevention nearly impossible. Hindu-Muslim rioting has caused the deaths of thousands of civilians, mainly Muslim but also Hindu, throughout northern and central India. The exception in the north is Punjab, only because all Punjabi Muslims migrated to the Pakistani section of Punjab (and Hindus and Sikhs from Pakistani Punjab into Indian Punjab) after the partition of that province in 1947. The Dravidian

TABLE 3.1
Communal Violence and Casualties in India

Year	*Incidents*	*Persons Killed*	*Persons Injured*
1974	248	87	1,266
1975	205	33	972
1976	169	39	794
1977	188	36	1,122
1978	230	110	1,853
1979	304	261	2,379
1980	427	375	2,838
1981	319	196	3,025

SOURCE: *Report: 1979–80*, Ministry of Home Affairs, Government of India, 1980, p. 5; and *Report: 1982–83*, Ministry of Home Affairs, Government of India, 1983, p 3.

NOTE: In 1969, over 1,000 deaths occurred due to communal rioting in Gujarat. Communal rioting caused more than 5,000 deaths in 1983 in Assam and more than 1,000 deaths in 1984 in Maharashtra. Official national statistics of communal violence for these years are not available

South (with the exception of areas that are now Andhra Pradesh, which was once ruled by the Nizams of Hyderabad) historically has remained outside the main Muslim conquests and Hindu-Muslim struggles, and since Independence communal violence there has been less widespread and intense than elsewhere. The extent of communal violence in India may be seen in the year 1979–1980, when such violence occurred in Jamshedpur and Purnea (Bihar), Aligarh (Uttar Pradesh), Nadia (West Bengal), Ahmedabad (Gujerat), Hyderabad (Andhra Pradesh), Chamarajpet (Karnataka), and Palacode (Tamil Nadu).[25]

Communal violence and bloodshed have been steadily on the rise since the mid-1970s (see table 3.1).[26] Pakistani and other critics of India have alleged that this is an indication that India's claim to be a secular state is a farce. However, the increase is not so much due to the failure of secular policy as it is to the increasing assertiveness of Indian Muslims for a greater share of economic and political power. This has been felt especially at the lower and mass levels of Indian society and has produced resistance by Hindus, who fear the loss of benefits from Indian development programs.

[25] See *Report: 1979–80*, Ministry of Home Affairs, pp. 5–9.

[26] For a brief discussion of communal and caste conflict in India, see Hardgrave, *India: Government and Politics*, pp. 120–27.

For more than two decades after the creation of Pakistan as a homeland for Indian Muslims in 1947, the large Indian Muslim population left behind in India found themselves without a strong and broadly based leadership. Most of the elite Muslim leaders had moved to Pakistan. Indian Muslims were also confused about their national identity and loyalty after the partition of India and for the most part were less visible in public life. However, the generation of Indian Muslims that grew up believing that there were two nations in India of Hindus and Muslims has since passed from the political and economic scene. Following the breakup of Pakistan in 1971, Mohammed Ali Jinnah's two-nation theory based on religion alone became less meaningful among Indian Muslims. Hindu-Muslim conflict in India may not, therefore, suggest the growing alienation of Muslims but their increasing identification with India and greater participation in the mainstream of Indian society.

A similar problem may also be seen in the increasing conflict and bloodshed between caste Hindus and Harijans. Harijans are the former "untouchable" castes, who are officially classified as the "Scheduled Castes" in reference to the Schedule of the Constitution that grants them special benefits and privileges. More than thirty years of special programs has produced an assertiveness among the Harijans, who numbered more than 100 million people in 1980.[27] Their progress in itself has produced resentment and scattered resistance and violence. Increasing communal and caste conflict in India may also be attributed to the growing demands by several religious, linguistic, and social groups for a larger share of the economic pie. The spark of a communal riot in India is often a trivial incident or rumor. But once passions are inflamed, the violence is carried out by what the government calls small groups of antisocial elements and by what other observers have perceived to be mostly "frustrated young people whose expectations have been awakened by the opening of previously closed post-colonial societies."[28]

[27] The figure of 100 million Harijans is obtained from ibid., p. 124. There are no precise figures on the actual number of former "untouchables." One government report roughly placed the number of "Scheduled Castes and Tribes" at around 22 percent of the Indian population. Assuming that about 5 percent are tribals, this would suggest that the Harijan population of India in 1984 would be about 150 million in a total population of 730 million. See section on "Scheduled and Backward Classes" in *India 1981*, pp. 117–28.

[28] William K. Stevens, "There is No End to the Subcontinent's Subdivisions," *New York Times*, May 13, 1984. According to Mrs. Gandhi, communal violence is inevitable in a country as varied as India, with its multitude of social and economic problems.

TABLE 3.2
Labor Unrest and Violence in India

Year	*Incidents*	*Percentage of Violent Incidents*
1977	6,258	11.6
1978	6,666	12.6
1979	4,353	15.7
1980	3,028	17.9
1981	4,182	18.9
1982	3,690	17.6

SOURCE: *Report: 1982–83*, Ministry of Home Affairs, Government of India, 1983, p. 4.

NOTE: A "violent incident" in this context involves injury to persons or damage to property due to clashes among civilians or between civilians and law-enforcing authorities

There have been other cases of national civil disobedience accompanied by occasional violence to protest economic and social conditions and the performance of the government in power. Labor unrest and calls for citywide or nationwide strikes often deteriorate into pitched battles between strikers and police. (See table 3.2.) In some cases, the protest movements are not directly related to trade unions but are called by opposition parties who want to embarrass or topple the government because of their perceived economic and political failures in office. Since this has proved difficult to achieve via the ballot, the strategy has been to resort to civil disobedience.

The protests and riots of 1974–1975 over higher prices and the scarcity of goods following the international oil crisis and monsoon failures provide a good example. The political demonstrations and call for civil disobedience in 1975 were mainly organized by opposition parties that included the Socialist party, the Jan Sangh, the Bharatiya Lok Dal and the Congress (Organizational) party in the Lok Sabha led by the elder socialist statesman, Jaya Prakash Narayan. Protest marches and various disturbances eventually led to the declaration of the National Emergency and the suspension of fundamental rights in June 1975. As in all such political demonstrations arising from economic, social, and ideological issues, the government's response has been to use the local police, the state-armed constabulary, and occasionally paramilitary forces. In some instances, the army has been called in.

See also William K. Stevens, "Indian Zealots Test the Limits of Democracy," *New York Times*, June 17, 1984.

II. The Internal Use of Force

The Instruments of Force

Four types of organizations in India use force in the maintenance of internal security. These are the local police, usually armed only with *lathis* (long sticks) and batons; the state-armed constabulary (the State Armed Police), often carrying outmoded rifles and bayonets; paramilitary forces, including the Home Guards, usually armed with more modern rifles and operating with broader objectives and functions; and the armed services, which nearly always means the Indian army, although the navy has occasionally been put on alert to guard installations and material on the docks that might be destroyed during civil disturbances. The police and the armed constabulary are under the jurisdiction of the state governments; the paramilitary forces are mainly controlled by the Ministry of Home Affairs; and the armed forces by the Ministry of Defence.

The use of these organizations tends to reflect the level and type of violence that the central, state, or local government is trying to deal with. Most violence—industrial strikes, protest demonstrations, and minor and brief communal riots—is dealt with by local police constables commanded by police inspectors armed with pistols. When the disturbances become more violent, the State Armed Police and, if necessary, the Home Guards are brought in. If these disturbances do not last long and provided there are relatively few casualties, the use of force at the first two levels produces comparatively minor domestic political repercussions. All decisions regarding the deployment of force up to this level are made by the state governments that control these forces. When the violence is intense, sustained, and results in extensive casualties, paramilitary forces are brought in. Failure to stop the violence at this level then produces military intervention. Because paramilitary forces and the armed services are under the Ministries of Home Affairs and Defence, the decision to deploy these forces is made by the central government.

The armed services under the Ministry of Defence are intended basically for external defense; their use in the maintenance of internal security is a measure of last resort. Whether they should be used at all is an issue that will be discussed in the next section. The other internal security forces under the central government are controlled mainly by the Ministry of Home Affairs but also the Ministries of Defence, Railways, and Shipping and Transport. These forces are considered to be paramilitary and exist between the State Armed Police under state governments and the armed services under the Ministry of Defence. Lt. Gen. M. L. Chibber has grouped the forces that are deployed to maintain internal security into three functional categories and the discussion that follows is based on his divisions.[29]

The first category is made up of those forces whose primary responsibility is to guard the national borders. They include the Border Security Force (BSF), the Indo-Tibetan Border Police (ITBP), the Uttar Pradesh Special Police Force (UPSPF), the Assam Rifles, and the Coast Guard (a wing of the Indian navy under the Ministry of Defence.) Each was initially raised to meet a specific border security problem.

The most prominent of these forces, the BSF, was raised after the Pakistani military incursion into the Rann of Kutch in April 1965. The chief of army staff at that time, Gen. J. N. Chaudhuri, proposed a lightly armed paramilitary force that would patrol the border with Pakistan in peacetime and be relieved by the army as soon as war broke out or was imminent. The BSF was formed in the same year and is now deployed along the Indo-Pakistan, Indo-Burma, and Indo-Bangladesh borders. It is armed with both small arms and a complement of heavy weapons. An air transport wing has been added, equipped with Indian-made HS-748 aircraft and some small imported aircraft such as the Super King Air Beach. The force is commanded by officers who have risen through the Indian Police Service cadres and by a few retired army officers. It has increased in size from about 55,000 men in 1968 to about 75,000 men in 1974 and has remained at that figure for the past decade.[30] According to the 1982–1983 report of the Ministry of Home Affairs, the BSF also provided assistance to civil administration in law and order duties in West Bengal, Assam, Maharashtra, Haryana, and Punjab. These functions are primarily the responsibility of the local police forces and the Central Reserve Police Force (see below). The BSF has also been deployed to protect the oil pipeline link between Assam and Bihar and in counterinsurgency operations in Tripura and Mizoram.

[29] Lt. Gen. M. L. Chibber, *Paramilitary Forces* (New Delhi: United Services Institution of India, 1979). See also K. P. Misra, "Paramilitary Forces in India," *Armed Forces and Society* 6, no. 3 (Spring 1980):371–88; and the assessment and recommendations of the former director-general of the Border Security Force, K. F. Rustamji, "India's Paramilitary Forces," *Institute for Defence Studies and Analyses Journal* 10, no. 1 (July-Sept., 1977):1–12.

[30] The various strengths of the paramilitary forces under the Ministry of Home Affairs are derived from its *Report: 1979–80* and *Report: 1982–83*.

The ITBP, originally conceived after the Sino-Indian treaty was signed in 1954, was intended to regulate trade and other traffic along the Indo-Tibetan border in the Northeast Frontier Agency (now called Arunachal Pradesh) in accordance with the new treaty. After the war with China in 1962, the role of the ITBP became similar to that described for the BSF. This force had a strength of nine battalions in 1983. The UPSF was also raised after the Sino-Indian war and patrols the Uttar Pradesh–Tibet border.

The oldest of the border forces is the Assam Rifles, founded by the British more than a century ago. It existed under various names such as the Assam Frontier Police, the Eastern Bengal, and the Assam Military Police until it received its present name in 1917. After Independence, it was initially placed under the Ministry of External Affairs and was used to police the tribals in the border areas. After the 1965 Indo-Pakistani war, it was moved to the Ministry of Home Affairs. A significant characteristic of the Assam Rifles is that it is in many ways trained and equipped as a regular infantry regiment with its officer corps provided by the Indian army. In fact, the bulk of the force continues to operate in the insurgency areas of northeast India under the army's operational control. In 1983, the Assam Rifles comprised twenty-one battalions, one training center, and other small ancillary units such as a construction company, a signal unit, and a base administrative organization.

The Coast Guard was established in 1976 and is the only border patrol force to be placed outside the Ministry of Home Affairs and under the Ministry of Defence. The Coast Guard was a response to the rapid growth of India's shipping and seaborne trade and the extension of territorial waters to two hundred nautical miles. It patrols and regulates commercial traffic within India's territorial waters. As in the relationship of the land border security forces to the Indian army, the Coast Guard during times of war is expected to function under the Indian navy and to undertake tasks such as surveillance, patrolling, and guarding harbors.

The second category of paramilitary forces consists of those intended to maintain internal law and order. These forces usually supplement the functions of the regular local police force and include the Central Reserve Police Force (CRPF), Central Industrial Security Force (CISF), the Railway Protection Force (RPF), the Defence Security Corps (DSC), the Home Guards, and the State Armed Police (SAP). Following the controversial deployment of the Indian Army in the Punjab in 1984, and in order to minimize the role of the army in maintaining internal security in the future, a new elite paramilitary force called the National Security Guards (NSG) was raised in 1985. Beginning with a modest size of 5,000 men, the NSG was expected to grow further by drawing mainly from the officers and enlisted men of the Indian Army after the early completion of their commissions.

Of the above forces, the most politically significant is the CRPF. It was established in 1940 during British rule and has been sent into action by the central government in New Delhi on several occasions in areas where the state governments have been unable to maintain law and order. It has been used to control industrial strikes and political demonstrations in Maharashtra, Assam, West Bengal, and Punjab as well as to fight insurgency in the border areas of Manipur and Mizoram. In a sense, the roles of the CRPF and the BSF have overlapped; both forces have been used to deal with problems that appear to fall within the functions of the other—the CRPF with border security and the BSF with internal law and order.

The growth of the CRPF has been steady, rising from 870 men in 1947 to 12,045 in 1962.[31] Within two years after the October 1962 Sino-Indian war, the strength of the CRPF had increased to 25,015. Subsequently there were major increases and the force reached a peak of 76,350 men in 1975, the year that the Emergency was declared by Prime Minister Indira Gandhi. The use of the CRPF during the Emergency was disputed by opposition party leaders. With the victory of the Janata government in the national elections in 1977, the size of the CRPF began to be reduced; it numbered 73,712 men in 1979. Controversy still surrounded the CRPF in 1980, when it was further reduced to 65,495 personnel.

The CISF was raised under an act of Parliament in 1968 and the personnel and deployment of the force continue to be governed by the provisions of the act. The CISF provides security at public sector undertakings. Its growth has been quite rapid, rising from 6,062 initial personnel to 39,660 in 1980, deployed at 100 government-run industrial corporations. Three years later, the size of the CISF had increased to 49,085 men and officers, who were assigned to some 115 corporations. Similar to the CISF was the Railway Protection Force, under the Ministry of Railways. As its name suggests, the RPF was developed in order to protect railway lines and other properties of the Indian Railways. It was established on a statutory basis in 1957 although it had antecedents in British times dating back to the 1880s. The RPF is comparatively large, comprising about 64,000

[31] See *Report: 1979–80*, Ministry of Home Affairs, p. 75.

personnel in 1980, including a subdivision known as the Railway Protection Special Force that consisted of eight battalions of well-armed men. Although about half of of the regular RPF carry ordinary rifles, the special force carries machine guns and other light weapons, signifying the importance that the government attributes to the Indian railway network for the maintenance of national security. The DSC, correspondingly, exists to protect army, air force, and naval installations in rear bases and ordnance factories under the Ministry of Defence.

The largest of the internal law and order forces is the Home Guards, an organization that existed in some states before Independence. The Home Guards were greatly expanded after the 1962 Sino-Indian war and by 1983, the size of the Home Guards stood at 470,000. This force may be considered most similar to the Peoples' Liberation Army in China; it performs a variety of functions ranging from civil defense to socioeconomic development assistance. The Home Guards is a voluntary force raised by state governments and Union Territory administrations but remains under the overall supervision of the Home Ministry. It is intended to be used by state governments as an auxiliary force for the maintenance of law and order and may also be called in for emergency assistance during natural disasters such as cyclones, floods, and epidemics, and to provide special functions such as air raid management and first aid. Since the late 1970s, the force has been used to promote communal harmony and to protect underprivileged classes in society, as during attacks on former "untouchables" in the villages by high-caste Hindus. In 1979, some states used the force for the promotion of socioeconomic and welfare activities such as adult education, health and hygiene projects, and other development schemes. Home Guards assigned to these functions are unarmed. However, the basic role of the Home Guards remains one of civil defense and for this they are appropriately armed, trained, and formed into special battalions. In the early 1980s, up to 30 percent of the Home Guards in the Indian states along the international borders were armed, and up to 20 percent in the other states.

Finally, the State Armed Police deal with all forms of violence that the local constabulary cannot handle and that do not appear to call for intervention by the centrally controlled paramilitary forces. SAP remains essentially a supplement to local police forces and is organized to deal with problems on a statewide basis.

Paramilitary forces in Chibber's third functional category are those that contribute to the nation's development. The main force in this category with a noticeable military link is the Border Roads Organisation (BRO). This force belongs to neither the Ministry of Home Affairs nor the Ministry of Defence but is under the Ministry of Shipping and Transport. The BRO was set up in 1960 for the purpose of developing roads and other lines of communication in strategic border areas and is staffed with a mixture of civilian and army engineers and administrators. A sizable number of army officers command key positions in this organization. The Home Guards also perform some developmental functions and thus would fall within this category as well.

The statutory basis for the establishment and development of all these paramilitary forces is found in Article 355 of the Indian constitution, which states that "it shall be the duty of the Union to protect every state against external aggression and internal disturbances and to ensure that the government of every state is carried on in accordance with the provisions of this Constitution." These forces were created to minimize the need to bring in the armed forces for the maintenance of law and order.[32] Under the Ministry of Home Affairs or other central ministries, they provide the central government with an alternative resort to maintain law and order in areas where state governments are perceived to have failed.

The Military and Internal Security

Controversy over the internal use of force surrounds both the paramilitary forces and the armed services. In the case of paramilitary forces, the issue is not whether such forces are needed for the maintenance of internal security but whether these forces are likely to be misused by the government in power. The rapid growth of these forces also suggests the increasing power of the central government to usurp the primary responsibility granted to the state governments by the constitution to maintain law and order. There are also problems regarding the role originally intended for these forces and their actual use. In the case of the armed services, the question is whether such forces should be used at all in the maintenance of internal security.

Controversy regarding the paramilitary forces was particularly high during the Emergency period between June 1975 and March 1977, when they were widely used under the 42nd Amendment that was passed by Indira Gandhi's Congress government in 1976. This amendment, which added Article 257-A to the constitution, states

[32] See S. Sahay, "Military and Paramilitary Forces," *Statesman*, August 28, 1980.

that the "Government of India may deploy any armed force of the Union or any other force subject to the control of the Union for dealing with any grave situation of law and order in any state."[33] Moreover, in deploying these forces, whether paramilitary or military, the central government would not "be subject to the superintendence or control of the state government or any officer or authority subordinate to the state government." The central government could thus deploy armed forces under its control anywhere throughout India without the consent of the state governments. Even the existence of such manpower in the hands of the central government could imply considerable coercive power to compel the states, especially those controlled by non-Congress governments, to follow the central government's directives.

The 42nd Amendment was immediately repealed by the Janata government, which assumed power in March 1977 following the defeat of the Congress party at the national elections. The new government passed the 44th Amendment in 1978 and once again made the deployment of paramilitary forces dependent on the consent of the state government. The repeal of the 42nd Amendment has not subdued the controversy over the internal use of force by the central government, however. Paramilitary forces have continued to be used all over the country for a variety of reasons, especially where state governments have collapsed and President's Rule has been imposed. Examples in the early 1980s are Assam and Punjab.

A further controversy concerns the plethora of border security forces under various titles. There has been some question whether they should be merged under a centrally commanded border security force. This would in effect separate paramilitary forces intended for the maintenance of external security from those intended for the maintenance of internal security. Army commentators such as Maj. Gen. K. S. Bajwa insist that there should be a clear demarcation of such roles.[34] According to Bajwa, "while the ethos of the former will be organized armed power, sustained vigilance and quick responses, the philosophy of the latter must be to dissipate violent protests and diffuse problems by persuasion and the use of minimum force." Gen. Bajwa has proposed that border security forces should work in conjunction with the armed services, while paramilitary forces for the maintenance of internal security

[33] See K. P. Misra, "Paramilitary Forces in India," p. 387n.

[34] Maj. Gen. K. S. Bajwa, "Para-Military Forces: Need for Clear Division of Duties and Command," *Indian Express*, August 28, 1980.

should be a central government extension of the state police. Forces in the first category would come under the jurisdiction of the Ministry of Defence; forces in the second would be under the Ministry of Home Affairs.

An even more serious issue regarding the use of force has been the use of the armed services to maintain internal security. (See table 3.3.) Almost to a man, the leaders of the armed forces are convinced that this should be avoided altogether, for three principal reasons.[35] First, the armed services constitute one of the chief unifying forces of the Indian Union. They are perceived by the Indian public as the defenders of their freedom. To use the armed services against their own people would produce a breakdown in the military-civilian trust that has been built up over three decades. Second, the armed services are not equipped or trained to deal with problems of internal law and order. As Lt. Gen. M. L. Thapan noted, "a fundamental principle of war is concentration of men and materiel at the right place and at the right time."[36] Internal security duties, on the other hand, "require dispersion and the use of minimum force since our own countrymen are involved." Lt. Gen. A. M. Vohra comments that because the army does not mingle with the crowd as the police are required to, the army's ability to sense and deal with internal riots is severely limited.[37]

Third, the use of the armed services for internal security may produce a breakdown in military training and readiness for dealing with external security. Their politicization and corruption might also be encouraged. Frequent use of the military for internal purposes would invariably arouse the emotions of both soldiers and civilians and could eventually result in struggles between the two sides leading to military coups and takeovers.[38]

Despite military opposition, civilian leaders feel that situations may sometimes exist in which limited military participation becomes inevitable even if undesirable. In an interview in 1981, Jagjivan Ram, former defense minister in the Congress government and

[35] Interviews conducted in 1980–1981 with former chiefs of air force staff, Air Chief Marshals Arjan Singh and P. C. Lal; former chiefs of naval staff, Admirals A. K. Chatterji and S. M. Nanda; and former vice chief of army staff, Lt. Gen. A. M. Vohra. Similar views may be found in the writings of Gen. J. N. Chaudhuri, Lt. Gen. M. L. Thapan, Maj. Gen. D. K. Palit, and Adm. S. M. Kohli.

[36] Lt. Gen. M. L. Thapan, "The Army: How Far to Aid Civil Authority," *Statesman*, August 3, 1980.

[37] Lt. Gen. A. M. Vohra, "Other Tasks of Armed Forces," *Tribune*, April 14, 1980.

[38] Lt. Gen. A. M. Vohra, "Coups and Military Take-Overs," *Tribune*, March 28, 1980.

TABLE 3.3
Army Deployment in Support of Civil Authorities, 1973–1985

Year	Location	Cause	Duration
1973	Assam	language riots	1 month
	Lucknow, U.P.	police unrest	3 weeks
	Arunachal Pradesh	tribal violence	1 week
	Imphal, Manipur	riots	1 week
	Nasik, Mah.	electoral violence	2 days
1974	Baroda, Guj.	communal violence	1 week
	Ahmedabad, Guj.	food riots	3 weeks
	Dhanbad, Ranchi, Bihar	communal disorder	2 weeks
	Patna, Bihar	student riots	1 week
	all India	national rail strike	3 weeks
	W. Bengal, 4 districts	riots	2 weeks
1975	all Indian port cities	dockers' strike	1 week?
	Delhi	communal riots	1 week?
	Cooch Behar, W.B.	rural unrest	1 week
1976	State of National Emergency declared; army not called out in aid to the civil authorities		
1977–79	No army utilization mentioned during Janata years in Ministry of Defence annual reports except for aid to the civil authorities in cases of natural disaster		
1980	Assam	"anti-foreigner" protests and election violence	continuous
	Tripura[a]	tribal violence	6 months
	Meghalaya[a]	tribal insurgency	continuous
	Manipur[a]	student/tribal unrest	1 month
	Nagaland[a]	tribal violence	continuous
	U.P., M.P., Gujarat, Delhi, and elsewhere?	communal violence	2 months
	Bihar	CISF strike	1 week?
1981	Assam	"anti-foreigner" protests	1 year
	Gujarat	protests over job quotas for Harijans	3 months
1982	Assam	"anti-foreigner" protests	1 year
	Arunachal Pradesh	student/tribal unrest	4 weeks
	Goa	communal disturbances	5 days
	Baroda, Guj.	communal disturbances	2 weeks
	Kerala	communal disturbances	4 days
	Maharashtra	Bombay police strike	1 month
	Mizoram, Nagaland[a]	election violence	4 days?
1983	No known army deployments other than continuous counterinsurgency operations in northeastern states and peacekeeping duties in Assam; army on alert in several states (Punjab, Haryana)		

TABLE 3.3 (*cont.*)

Year	*Location*	*Cause*	*Duration*
1984	Maharashtra (Bombay and nearby suburbs)	communal riots	4 weeks
	Punjab, Chandigarh, and nearby areas	Sikh terrorist campaign; army mutinies	ongoing from June 5
	Hyderabad, A.P.	communal riots, political disorder	4 days?
	New Delhi and many other cities	anti-Sikh riots in wake of Mrs. Gandhi's assassination	1 week
1985	Continuing security deployments for domestic unrest in northeastern states, Assam, Punjab, and perhaps selected districts in Haryana and Jammu and Kashmir.		
	Ahmedabad, Guj.	caste riots over government job quotas for Harijans; communal violence	1 week
	Punjab, Haryana, New Delhi	Sikh terrorist campaign	1 week

SOURCE: Lt. Col M.M L Ahuja, "In Service of the People," *Sainik Samachar*, January 17, 1981, p. 46 (in Stephen P Cohen; "The Military and Constitutionalism in India," paper presented to the Conference on India's Democracy, Princeton University, March 1985), with some updating.

[a] The Indian army has been on continuous alert status and active in counterinsurgency and antiterrorist activities in these states for several years

deputy prime minister in the Janata government, stated that in the cases of extreme breakdown of law and order, the mere introduction of army units in troubled areas tends to produce a pacifying atmosphere without these units having to resort to any actual force. (See table 3.4.) Indeed, the army rarely acts when deployed in strife-torn areas and thus this reduces the deaths usually caused when police finally use their weapons. However, military leaders interviewed in 1981 pointed out that this was true only because the armed forces had thus far been used only sparingly. Frequent use, on the other hand, could destroy the pacifying effect, and consequently they feel that the goal of the central government should be the total avoidance of the use of the armed services for internal security.

However, in two domestic areas military leaders agree that the use of the armed services may prove unavoidable. The first concerns the problem of dealing with insurgencies, as in the northeast of India. But even here the military caution against its frequent or continued use. Persistent insurgencies like those in Nagaland or Mizoram, as Generals Vohra and Thapan pointed out, suggest the need for political settlements rather than military solutions.[39] The second

[39] Lt. Gen. A. M. Vohra, "The Anatomy of Insurgency," *Tribune*, May 23, 1980; and Thapan, "The Army: How Far to Aid Civil Authority."

TABLE 3.4
Reasons for Army Deployment in Support of Civil Authorities, June 1979–December 1980

	Law and Order	*Maintenance of Essential Services during Strikes*	*Disaster Relief*	*Other*[a]
Bihar	6	1	1	4
Gujarat	4	0	4	1
Punjab	2	2	0	5
Karnataka	0	0	3	1
Delhi	2	2	0	5
Haryana	0	0	2	1
Kerala	3	0	1	0
Orissa	3	0	1	3
Madhya Pradesh	3	0	2	3
Tamil Nadu	2	0	1	0
Andamans and Nicobar	2	0	0	0
Meghalaya	4	1	1	1
Assam	2	0	5	0
West Bengal	0	0	1	12
Manipur	2	0	1	0
Nagaland	2	0	0	11
Tripura	1	0	0	0
Uttar Pradesh	17	0	2	27
Jammu and Kashmir	7	0	0	40
Arunachal Pradesh	1	0	0	1
Mizoram	1	0	2	25
Himachal Pradesh	0	1	1	14
Maharashtra	0	0	14	2
Andhra Pradesh	0	0	2	1
Rajasthan	0	0	3	3
Sikkim	0	0	0	3
Totals	64	7	46	163

SOURCE: Lt. Col. M.M.L. Ahuja, "In Service of the People," *Sainik Samachar*, January 17, 1981, p. 46. (in Stephen P. Cohen, "The Military and Constitutionalism in India," paper presented to the Conference on India's Democracy, Princeton University, March 1985.

[a] The "Other" category includes mine detecting, bomb disposal, vehicle recovery, and road and bridge maintenance.

area where the military agree it has a contribution to make is in disaster relief. There is general consensus within the military that this constitutes one of their essential domestic tasks.[40]

The Indian army's misgivings over its use in the maintenance of internal security appear to have been justified. Although there have been several disturbances in the police and paramilitary forces, the army's assault on the Golden Temple in June 1984 set off the first, though brief and limited, military mutiny in independent India. (See table 3.5.) Only two other mutinies had previously occurred in the history of the Indian armed forces and both had occurred under the British. The first was the Great Indian Mutiny of 1857–1858 when Indian sepoys (foot soldiers) revolted against their commanding British officers; the second was the Indian naval mutiny by Indian seamen against their British naval commanders in 1946. In retrospect, both mutinies were seen as milestones in the road toward Indian independence.

The army assault on the Golden Temple triggered off Sikh mutinies and desertions from Sikh regiments at military encampments in Ramgarh (Bihar), Sri Ganganagar (Rajasthan), Poona (Maharashtra), and in the state of Tripura. At Ramgarh, six hundred Sikh troops deserted their unit and wounded six hundred people, including twenty-five officers, in a fierce battle.[41] The mutineers then fled with their arms in army trucks, buses, and cars; they vowed to retake the Golden Temple from the army. According to official Indian estimates, there were about one thousand desertions and almost all of these were raw recruits to the Sikh regiments.[42] Army spokesmen claimed that in the three-day Sikh mutiny, forty-three Sikh soldiers died in the army's efforts to suppress the revolt. (The Sikhs, who constitute 2 percent of the Indian population, constitute about 10 percent of the armed personnel in the Indian army of 1.2 million men.) Other government officials claimed that for the most part, Sikh soldiers remained loyal to their units; this was highlighted by the fact that the army assault was led by a Sikh, Lt. Gen. Rajeshwar Singh Dayal, together with several other Sikh soldiers under his command.

Nevertheless, the Sikh mutiny of 1984 raises fundamental questions about the domestic role of the armed forces. The short-term need after the army assault on the temple was to restore the holy

[40] See Vohra, "Other Tasks of the Armed Forces."

[41] *New York Times*, June 11, 1984.

[42] See *New York Times*, June 13, 1984; and *Indian Express*, June 12, 1984.

TABLE 3.5
Disturbances within Peacekeeping Forces since 1978

Year	*Location*	*Disturbance*	*Outcome*
1978	Tamil Nadu	CRPF strike; state police unwilling to move against strikers	2 batt. of BSF end strike
1979	Gujarat	police strike	army restores order; 5,000 arrested; army fires on striking policemen
1979	Bhubaneswar, Puri, Trivandrum, Cuttack, Neemuch, Thumba, Cochin Madras, Pt. Blair, Delhi	CRPF, CISF, RPF strikes	army and BSF disarm strikers; 24 CISF and 3 army jawans (enlisted men) killed at Bokaro; 3 CRPF mutineers killed in Delhi
1979	Tamil Nadu	police strike	BSF, CRPF arrest 3,000 police
1979	Bombay	abortive police strike	BSF, CRPF quell Cong.-supported strike
1980	various Bihar, West Bengal, industrial locations	CISF strike, clash with police	army disarms CISF
1981	Gujarat	police ineffective during protests against employment quotas for Harijans	army restores order
1982	Bombay	police strike	army restores peace
1984	Army barracks in Bihar, Maharashtra, Tripura, Rajasthan, elsewhere?	uncoordinated mutiny by 1,500 Sikh jawans in aftermath of army actions in Golden Temple	loyal army units capture mutineers; some flee to Pakistan? Under trial for mutiny?

SOURCE Lt. Col. M.M L. Ahuja, "In Service of the People," *Sainik Samachar*, January 17, 1981, p. 46. In Stephen P. Cohen, "The Military and Constitutionalism in India", paper presented at conference on India's Democracy, Princeton University, March 1985

NOTE: BSF–Border Security Force; CRPF–Central Reserve Police Force, CISF–Central Industrial Security Force; RPF–Railway Police Force.

shrine to Sikh civilian control under the Shiromani Gurdwara Prabandhak Committee (SGPC), the managing body of the temple, and to pacify the anger felt by almost all Sikhs throughout India and abroad. From a military standpoint, it was also important to restore the confidence and commitment of the army's Sikh soldiers, who have historically distinguished themselves through tremendous valor and sacrifice. Since the assassination of Mrs. Gandhi in October 1984 and the succession of Rajiv Gandhi as prime minister, these needs have been met.

In the long term, the solution may lie in separating (or shielding) the military from the role of maintaining internal security. This will require a clear delimitation of functions among the military, border security forces, and paramilitary forces. The military's role should be strictly in the realm of external security. Special counterinsurgency units may be deployed to deal with guerrilla warfare, particularly where this involves external military assistance to the guerrillas. But even here, the problem is basically internal and political and calls for appropriate solutions at these levels. Border security forces need to be utilized for what they were intended, the patrolling of borders and the protection of strategic targets in these areas. They should serve as a "trip wire" to provide warning of the potential onset of a wider military conflagration; regular military forces would then need to be called in. Paramilitary forces should be designed and trained for the maintenance of internal security and their use should be a measure of last resort when control by state police has failed. Despite fears of the increasing growth of paramilitary forces, their use may be preferable to the induction of Indian army units into riot-torn areas. The creation in 1985 of the National Security Guards, drawn mainly from the Indian Army but distinguished functionally and operated separately from the regular armed forces, may provide the well-trained and disciplined force needed to deal with severe acts of civilian violence within the country.

CHAPTER FOUR

The Political System and the Security Decision-Making Process

SECURITY policy making takes place within the domestic political milieu. A variety of historical and sociocultural factors shape perceptions of external and internal threats to the nation. In a broad sense, a nation's perceptions of threats may be seen as the accumulation of past experiences and prejudices forged through personal memories or historical records. It is difficult to understand Indo-Pakistani hostility without knowledge of Indian history and the nature of ethnic and religious antagonisms in India. The Sino-Indian border dispute may be seen as part of the British legacy bequeathed to the independent government of India. Similarly, explanations for various insurgencies and separatist movements in India may be found in the politics of the pre-Independence era.

More immediate political conditions, however, also shape Indian security policy. These include the structure of the political system within which decisions must be made and more specifically the constitutional powers that define the government's ability to formulate and conduct internal and external security policies. In discussing the political structure, this chapter is concerned with how the system operates within the framework of the constitution—the organization and behavior of political parties; the conduct of electoral and parliamentary politics; the manner in which standing and ad hoc parliamentary committees investigate and provide input to the policy-making process; and the suitability and adaptability of the parliamentary system of government for the making of security policy in India.

In the Indian constitution, the relevant provisions here are those that define the government's power over external and internal security problems, especially the emergency powers of the government during times of war or the internal breakdown of law and order. Unlike the controversies that surround the war powers of the executive and legislative branches of government in the United States, the controversy in India arises from the restrictions that are imposed on the democratic rights of citizens at home in order to conduct war with external states or to deal with internal threats. Indeed, there has been no disagreement thus far with the government's decisions to engage in external wars, other than questions regarding the adequacy of war preparations during such crises.

Beneath this broader political context, two other levels of political processes will be analyzed. At the second level, the national debate on security affairs and in particular the substance and quality of the debate in Parliament will be of interest. Although the political system is built on western parliamentary and federal models, the debate is shaped by the unique distribution of power in the Indian Parliament and the degree of knowledge on security matters that prevails both in and out of it. Finally, the third level of analysis will focus more narrowly on the governmental security decision-making process. This involves the interaction of three basic groups and their leaders: elected political leaders, primarily at the cabinet level; career civil servants, especially the secretaries of important departments; and the members of the armed services, especially the chiefs of staff of the army, air force, and navy.

1. The Context of Security Policy Making

Historical and Sociocultural Considerations

The relevance for national security of such intangibles as the historical experience and the sociocultural ethos of a state may not be readily apparent or easily defined.[1] Usually, the colonial experience and the struggle for independence are an important influence on the security policies of states in the Afro-Asian and developing world. In India this has implied, on the one hand, the absorption of British political and military structures, processes, and values. On the other hand, it has also led to the rejection—especially during the first fifteen years of independence—of the British and western approach to strategic planning and security policy formulation. The policies of nonalignment and *panchshil,* discussed in chapter one, may be interpreted as India's negative reactions to the western colonial experience.

Through much of the 1950s, Indian defense policies were also conditioned by the prevailing domestic political philosophy: the belief in Gandhian principles of nonviolence to achieve political ends. However, Mahatma Gandhi's strategy of nonviolence against British imperial power in India may have been successful only because

[1] For example, the historical growth and consolidation of the United States has some relevance for the formulation of current national security policy. For a collection of readings on the historical roots of U.S. foreign policy, see Armin Rappaport, ed., *Essays in American Diplomacy* (New York: Macmillan, 1967).

it was a domestic civilian movement against a somewhat restrained colonial power.[2] In the conduct of external security measures, the utility of such tactics is not all that clear. Although the Indian National Congress, led by Gandhi, Nehru, and Patel, refused to endorse the British decision to enter India into the Second World War, this was largely due to the British failure to consult the Congress before taking the decision, rather than due to disagreement on whether India should fight the Axis powers. Congress had not ruled out the use of force for defense purposes or for national consolidation and, indeed, India was engaged in military action in Kashmir and Hyderabad.

Nevertheless, the Gandhian philosophy that advocated nonviolent methods and moral power in order to overthrow the colonial masters appears to have spilled over into the immediate post-Independence era, especially with the passing of the belligerent nationalist, Patel. This philosophy is implicit in Nehru's idealistic vision that security was better achieved through professions of peaceful coexistence and diplomacy rather than through arms buildups—an approach to security that was manifest in the 1950s at least in India's policy toward China, if not toward Pakistan. Note, however, that Nehru's concept of peace was also conditioned by the high priority that the new government gave to economic development.

There is another aspect of history that is crucial to understanding India's post-Independence security policy. The struggle for independence involved a parallel struggle for the creation of the Islamic state of Pakistan for Indian Muslims. Pakistan was eventually born in the bloodbath of civilian violence between Muslims on the one hand and Hindus and Sikhs on the other. Its formation soon led to war between the two newly independent states of India and Pakistan over the status of Kashmir. Memories of the violent struggle between Hindus and Muslims in pre-Independence India have remained one of the central historical experiences that has obsessed political leaders and defense decision makers in both countries.[3]

[2] On Gandhi's philosophy and strategy of nonviolence, see Louis Fischer, *Gandhi: His Life and Message for the World* (New York: New American Library, 1954); Suzanne H. Rudolph and Lloyd I. Rudolph, *Gandhi: The Traditional Roots of Charisma* (Chicago: University of Chicago Press, 1983); Judith M. Brown, *Gandhi and Civil Disobedience* (Cambridge: Cambridge University Press, 1977); Francis G. Hutchins, *Gandhi and the Quit India Movement* (Cambridge: Harvard University Press, 1973); and William L. Shirer, *Gandhi: A Memoir* (New York: Simon and Schuster, 1979).

[3] There are several historical and political studies on the Hindu-Muslim struggle in British India. See in particular the relevant chapters in Stanley A. Wolpert, *A New History of India* (New York: Oxford University Press, 1982); Percival A. Spear, *A His-*

This, together with the large Muslim minority left in India after partition, has made the government of India particularly sensitive to Islamic issues. It is primarily the Muslim question that underlies India's domestic and international policy of "secularism," although its roots may also be found in the agnostic outlook of Nehru.[4] The need to implement secularism has had relevance for the maintenance of both internal and external security.

At the domestic level, secular policy was intended to integrate the Muslim minority, a population of 85 million in 1983, into the general political and social fabric of India. It was feared that failure to integrate would lead to the alienation and antagonism of the minority and that such conditions would result in more communal rioting, threatening the political stability of India. It was also feared that the Muslim minority, left isolated, might produce a "fifth column" that could undermine India through espionage and sabotage during times of Indo-Pakistani crisis.

At the international level, the policy of secularism was related to the disputed state of Kashmir. The integration of Kashmir, which had a Muslim majority, into the Indian Union soon became a primary symbol of the government's commitment to the establishment of India as a secular state. Eventually, this produced a rigid attitude regarding negotiations with Pakistan over Kashmir and a perpetual confrontation between India and Pakistan. Indian perception of a Pakistani threat to take Kashmir by force if the appropriate opportunity presented itself has increased the Indian need to take appropriate defense measures to deter or defend against such an attack. The Kashmir issue and the preoccupation with the threat posed by Pakistan, has in turn produced a further general sensitivity to Islamic causes abroad, especially the Palestinian issue. The desire to retain Kashmir as well as to thwart Pakistan's propaganda and its efforts to establish military links with Middle Eastern states has prompted the Indian government to cultivate strenuously the friendship of other Muslim states, Arab states in particular.

---

tory of India, vol. 2 (Baltimore: Penguin Books, 1965); C. M. Philips and Mary D. Wainwright, eds., *The Partition of India: Policies and Perspectives, 1935–1947* (Cambridge: M.I.T. Press, 1970); and Larry Collins and Dominique Lapierre, *Freedom at Midnight* (New York: Avon Books, 1976).

[4] On religion and politics in India, see Donald E. Smith, *India as a Secular State* (Princeton: Princeton University Press, 1963) and Donald E. Smith, ed., *South Asian Politics and Religion* (Princeton: Princeton University Press, 1966). See also Paul R. Brass, *Language, Religion and Politics in North India* (New York: Cambridge University Press, 1974).

The policy of secularism, however, was not only intended to resolve the problem of the Muslim minority. Other important religious minorities needed to be integrated politically into the Indian Union as well. These included, in 1983 population figures, 18 million Christians, 12 million Sikhs, 4 million Buddhists, 3 million Jains, and 100,000 Parsees, who play a critical role in India's economy. Despite some grievances, political protests, and demands at various times, the policy of secularism has drawn all these religious minorities into the mainstream of Indian political and economic life in a productive and profitable manner.

In addition to religious minorities, another historical factor to be considered is the variety of linguistic and tribal nationalistic movements that were carried over from the pre-Independence to the post-Independence era. Bengali nationalism among both Hindus and Muslims had sometimes overshadowed Hindu-Muslim rivalry and violence. The Muslim Bengali leader, H. S. Suhrawardy, who later became one of the prime ministers of Pakistan, had once asked for an independent Bengali state outside of the proposed new states of India and Pakistan. Similarly, Tamil nationalism and the demand for a separate and independent state of Dravidistan composed of all the peoples of southern India belonging to the Dravidian linguistic group, had begun in the 1920s in the anti-Brahmin campaigns among the Tamil-speaking people and paralleled the Indian nationalist movement during the interwar years. Tribal separatism among the Nagas, Mizos, Gharos, and others has its roots in the British delineation of these areas for separate administration from the rest of Bengal and Assam due to their greater economic and social backwardness. Problems here were aggravated by the fact that the British also allowed American Christian missionaries to preach and proselytize in this region, thus creating enclaves of Christian communities in the northeast.

The Political System

Apart from the historical and socioreligious experience, Indian security policy is more immediately shaped by the prevailing political structure and its processes.[5] Of concern here is whether the system and conduct of government in India is appropriate for dealing with problems of external and internal security. The focus is essentially

[5] For a discussion of how political structures affect defense policy making, see Warner R. Schilling, "The Politics of National Defense: Fiscal 1950," in Warner R. Schilling, Paul Y. Hammond, and Glenn H. Snyder, *Strategy, Politics and Defense Budgets* (New York: Columbia University Press, 1962), pp. 1–27.

on Parliament, the various ministries and departments, and their accompanying bureaucratic procedures. This is where national policies are formulated and executed; consequently, the nature and performance of the parliamentary system broadly affects the success or failure of India's external and internal security policies.

As in the British system, the supremacy of Parliament in all matters is the cornerstone of the Indian political system.[6] Parliamentary control over defense and security issues means that ultimate authority is not held by the cabinet, the government in power, or the majority of the members of the party; rather, it rests with Parliament as a whole. According to Article 79 of the constitution, Parliament consists of the president and the two central houses: the powerful lower house, known as the Lok Sabha (House of the People) and the somewhat advisory upper house known as the Rajya Sabha (the Council of States). A similar arrangement exists in the state legislative assemblies, the Vidhan Sabhas, although not all states have two houses. Responsibility for national security is formally vested in the president and his council of ministers acting in conjunction with the Lok Sabha and Rajya Sabha.

The adoption of the British parliamentary system of government in 1950 by the framers of the constitution was based on two primary considerations. First, there existed previous parliamentary experience in India in the decades before Independence. The independent constitution of India was in fact a modified version of the Government of India Act of 1935 enacted by the British. Second, the Westminster form of government was perceived to possess greater elasticity and strength than the American presidential form. In rejecting the American system, the majority of these who drafted the Indian constitution felt that the independence of the legislative and executive bodies would result in frequent deadlocks, particularly in situations in which the legislature was dominated by one party and the executive by another. As the English jurist Walter Bagehot noted, under the American constitution, "the executive becomes unfit for its name since it cannot execute what it decides on; the legislature is demoralized by liberty by making decisions of which others (and not itself) will suffer the effects."[7]

The British parliamentary system, where there is a fusion of the

[6] Some portions of this introductory section are adapted from my book, *The Defence of India: A Budgetary Perspective of Strategy and Politics* (New Delhi: Macmillan Co. of India, 1978; Columbia, Mo.: South Asia Books, 1978), pp. 66–70.

[7] Walter Bagehot, *The English Constitution*, World Classics (London: Oxford University Press, 1930), p. 15.

executive and legislative powers, was expected to produce greater responsiveness to the needs of the fledgling Indian Union, faced with the immense task of maintaining political stability and stimulating rapid economic development. Both objectives would be jeopardized if the security and stability of India were neglected or if issues could not be dealt with swiftly. Such problems were particularly acute at the time of Independence and partition in 1947. The period between 1947 and the adoption of the constitution in January 1950 witnessed massive communal rioting in which more than half a million Hindus, Muslims, and Sikhs were killed and many millions more were uprooted and displaced. War also broke out between India and Pakistan over Kashmir and limited military action had to be taken to incorporate the "princely" state of Hyderabad into the Indian Union. Amidst this turmoil, the "father of the nation," Mahatma Gandhi, was assassinated. Accordingly, by the Seventh Schedule of the constitution, the Union Parliament was given legislative jurisdiction over all aspects of security, including defense and defense production.[8]

Control by Parliament is exercised through parliamentary debate and special standing investigative committees, notably the Estimates Committee, the Public Accounts Committee, and the Committee on Public Undertakings. Other ad hoc committees may be set up from time to time to investigate various problems in the public sector as well as to examine general political, economic, and social issues. An important feature of this system is that although Parliament is supreme, for the most part it exercises control ex post facto.[9] The entire legislative body does not make policy, but may guide or influence it in certain directions through debate and investigation. Past errors and misguided defense and internal security policies may be sharply criticized, especially by opposition members, causing the government in power to adjust or change its policies. However, the effectiveness of the system depends to some extent on the party or parties in opposition being able to displace the ruling party at the next election should the electorate perceive the government's policies to be inadequate.

Thus, in the mechanics of this system, Parliament does not enjoy powers of initiative or veto as long as one party dominates the lower house. This situation has generally prevailed in India since Inde-

[8] *The Constitution of India*, List I, Union Government, Seventh Schedule.

[9] See L. M. Singhvi, ed., *Parliament and Administration in India* (New Delhi: Metropolitan Book Company, 1972), p. 62.

pendence under the Congress party, except for a brief period between 1977 and 1980 after the newly and hurriedly constituted Janata party defeated the Congress party in the March 1977 national elections. Unlike the clear and often more than two-thirds majority held by the Congress party under Nehru and Mrs. Gandhi, the Janata party carried only a simple majority in the Lok Sabha and faced a strong Congress party in opposition. The collapse of the Janata government and the subsequent breakup of the party, together with the return of Indira Gandhi's Congress party to power after the December 1979 national elections, once again produced Congress dominance. This dominance was increased further in December 1984 when the Congress party, led by Rajiv Gandhi, won nearly 80 percent of the seats in the Lok Sabha. These conditions have enabled the ruling Congress party to wield the responsibility of Parliament as a whole in security matters. Disapproval by the small and fragmented political opposition could always be disregarded as of little consequence. At the same time, opposition within the Congress party during censure motions and votes of no confidence was likely to be disciplined by the party whip. Dissenting members are made to align themselves with government policies and actions.

One-party dominance in Parliament has led to crises and failures that might have resulted in the collapse or defeat of governments in similar systems, such as those of Britain, Canada, or Australia. Yet in India the Congress government has managed to survive votes of no confidence with relative ease, despite some shattering military failures—such as the crushing military defeat at the hands of China in the fall of 1962 and the military retreat against the Pakistani attack in the Rann of Kutch in April 1965. Similarly, the Congress government's decision to hold elections in Assam that triggered the massacre of Bengali Muslims in February 1983 and the decision in June 1984 to order the army to attack the Golden Temple are examples of internal crises of major proportions that failed to bring down the Congress government in power. Opposition members may shout or stage a walkout, but their ability to influence government policy appears to be limited.

However, the problem of one-party dominance and the inability of opposition parties to exert pressure on the ruling party either through voting in Parliament or through the potential to displace the Congress government in the next general election has been mitigated via other conditions and mechanisms. Prime Minister Nehru, for instance, was careful to consult opposition leaders on important policy issues despite their weak and fragmented representation in

Parliament; he governed through near parliamentary consensus. This practice, however, was much less visible under Congress prime minister Indira Gandhi or Janata prime minister Morarji Desai because the relationship between the ruling and opposition parties has tended to be more confrontational than cooperative. The situation under Prime Minister Rajiv Gandhi in 1985 is one in which the Congress party controls approximately 80 percent of the seats—the highest percentage in the history of independent India—and therefore opposition input into parliamentary debate is likely to be weak and insignificant.

Yet the ineffectiveness of opposition parties to influence government policies is offset by opposition within the ruling party itself. The Congress party of India over the years has been basically a "catch-all" party. Its members' ideologies range from the extreme left (strongly socialist and even some with Communist ties) to the extreme right (wealthy landowners, advocates of private capitalistic entrepreneurship, and hard-line religious communalists). Before 1975, the left-wing faction of the party influenced the shape of domestic and foreign public policies; since then, the centrist and right-wing factions have had greater influence. Although all Congress Members of Parliament may vote together to prevent the collapse of the central government, proposed Congress policies, programs, or legislation are likely to be compromises that attempt to satisfy the more important factions within the party.

Sometimes differences among important leaders within the party prove to be irreconcilable, leading to weak or ambiguous policies. A good example in the case of defense would be the situation between 1956 and 1962 when the left-wing V. K. Krishna Menon was defense minister and the right-wing Morarji Desai was finance minister. This incompatible relationship produced friction over the allocation of funds to the Defence Ministry and was partially responsible for the neglect of Indian defense capabilities that led to the Indian military defeat in the 1962 Sino-Indian war. Nothing like the Menon-Desai intraparty struggle for power occurred while Indira Gandhi was prime minister. She dominated the party and the government; threats to her leadership or destabilizing leadership rivalries within her cabinet led to the displacement of those involved.

The most significant input that the opposition has into the government's policies and programs is provided through certain parliamentary committee mechanisms that bring together the ruling party and opposition party Members of Parliament. Although the representation on these committees is usually in proportion to party representation in Parliament, opposition party M.P.s are able to influence the government's policies on various issues more directly and substantively than during the general debate in Parliament. The special investigative committees, notably the Estimates Committee, the Public Accounts Committee, and the Committee on Public Undertakings, periodically conduct extensive investigations into defense expenditures and into the workings of the defense industries.

A second group of these bodies that enhances the interaction of government and opposition may be found in the parliamentary consultative committees that are attached to every ministry. The committees are constituted by the Department of Parliamentary Affairs and consist of members from both the Lok Sabha and the Rajya Sabha. Their purpose is "to promote contacts between members of Parliament, Ministers and civil servants so as to acquaint the M.P.s through discussion with the principles, problems and working of government policies and public administration."[10]

The origin of these committees goes back to the Standing Committees first constituted in 1922 under the Montague-Chelmsford reforms. These committees continued to function until 1952, when they were abolished. In 1954, informal consultative committees were established through which the ministers met with M.P.s regarding policies under their ministries. In 1969, the informal consultations were formally constituted as the parliamentary consultative committees to function on a regular basis. The committees were reconstituted in June 1982 in order to bring new departments and ministries into existing or additional committees. There are now twenty-five parliamentary consultative committees, including committees attached to the Department of Atomic Energy and other scientific departments, the Ministry of Energy, the Ministry of Defence, the Ministry of External Affairs, and the Ministry of Home Affairs.

There are limitations on the influence of these committees. According to the guidelines formulated in 1969, "the deliberations of these committees would remain informal and no reference to the discussions held in the meeting thereof would be made on the floor of the House."[11] Moreover, the guidelines stipulate that "where

[10] *Report: 1982–83* (New Delhi: Dept. of Parliamentary Affairs, Government of India, 1983), p. 8.

[11] For this and the following quote, see "Appendix III: Guidelines Formulated in April, 1969 to Regulate the Constitution and Functioning of the Consultative Com-

there is unanimity of view in the Committee, Government will normally accept the view subject to the following exceptions: (i) any view having financial implications; (ii) any view concerning security, Defence, External Affairs and Atomic Energy; and (iii) any matter falling within the purview of an autonomous corporation." However, if the government fails to accept a consultative committee's unanimous or majority recommendation, it is expected to provide the reasons for the rejection.

Despite the above compensatory mechanisms and conditions, there have been misgivings regarding the effectiveness of the parliamentary system in India. Most of the doubts surfaced following the declaration of the Emergency in 1975. Although those who drafted the constitution between 1947 and 1950 reached a majority decision that a parliamentary system was preferable to a presidential system in India, there are now some Indian politicians and political scientists who feel that the presidential system may be more desirable for the maintenance of external and especially internal security.[12] However, although some Congress ministers such as C. M. Stephen, Kamalapati Tripathi, and Bhagwat Jha Azad have called for a close examination of the presidential system, former Janata foreign minister, A. B. Vajpayee, now of the Bharatiya Jạnata party, and former Janata prime minister, Morarji Desai, have rejected such Congress suggestions.[13]

The two principal reasons advocated for the adoption of a presidential system are first that it would enable an elected president, who would have a fixed period of tenure, and unelected cabinet ministers, who would not have to worry about reelection, to adopt policies that might be unpopular but effective in dealing with external and internal security problems. Second, the presidential system would resolve a basic problem that has plagued state governments. This is the practice of "floor crossings," in which members of the state legislative assemblies cross over to the opposition or ruling government benches. When this practice began on a major scale in

---

mittees for Various Ministries/Departments," in *Report: 1982–83*, Dept. of Parliamentary Affairs, pp. 32–33.

[12] See R. Chakraborty, "Do the People Want Change in the System?" *Statesman*, December 13, 1980; Justice A. K. Das, "Changes to Make the Government Work," *Statesman*, December 18, 1980; Nani A. Palkhivala, "Presidential System: A Question of Timing," *Hindustan Times*, January 7, 1981; Salman Kurshid, "Palkhivala's Volte Face," *Hindustan Times*, January 13, 1981.

[13] See *Sunday Standard*, December 7, 1980, *Statesman*, December 6 and 25, 1980; and *Hindustan Times*, December 28, 1980.

the late 1960s, toppling some state governments, it was discovered that no constitutional provisions prevented it. As a consequence, disgruntled members of the legislative assemblies of the majority party or coalition ruling party who were dissatisfied over not receiving ministerial posts or other expected rewards could with one swift motion bring the government down. This problem was resolved in 1985 when Prime Minister Rajiv Gandhi introduced an anti-defection law that deprived legislators of their membership in Parliament or the state assemblies if they quit their party to join another or if they were expelled from their party in accordance with parliamentary procedures. However, party splits or mergers were permitted provided they included one-third of the party legislators in the case of splits or received two-thirds approval from the party in the case of mergers.

In early June 1981, Mrs. Gandhi temporarily laid this controversy to rest by declaring that she had no intention of changing the system of government under her. However, this basic question of the suitability of the existing system of government to deal with acute crises will no doubt surface again in the future.

The Constitutional Powers

Although the broader political structure sets the boundaries within which policy decisions are made, various provisions of the constitution more specifically set the legal limits of policy making and enforcement. The most significant and controversial aspect of the Indian constitution pertains to the "emergency" powers granted to the central government to deal with problems of both external and internal security. Ironically, the emergency powers incorporated in the democratic Indian constitution were the legacy of British imperial rule. The Indian president's ability to proclaim a state of emergency when the nation's security was threatened was a power held by the British viceroy under the Government of India Act of 1935.[14] The British had exercised these emergency powers between 1940 and 1945 when they were at war with Germany and Japan and simultaneously had to deal with the Indian nationalist struggle for independence and with Hindu-Muslim violence.[15] Such viceregal powers were reinforced by the Defence of India Act of 1939, through which the state could resort to the "preventive detention"

[14] See Henry C. Hart, ed., *Indira Gandhi's India* (Boulder, Colo.: Westview Press, 1976), pp. 16–17.

[15] See Robert Frykenberg, "The Last Emergency of the Raj," in *Indira Gandhi's India*, ed. Hart, pp. 37–66.

of those not charged with committing criminal acts but who threatened internal law and order for political reasons. There are basically three categories of emergencies as defined by the constitution.

National Security Emergency. Under Article 352 of the constitution, the president, acting on the advice of the prime minister and his cabinet, may proclaim a state of emergency if the security of India or any part of its territory is threatened by external aggression or internal disturbance.[16] The proclamation may be issued by the president in anticipation of external aggression or internal disturbance, provided he is satisfied that the danger is imminent. However, the proclamation must be approved by Parliament within two months or it becomes ineffective. Once the emergency is in effect, Articles 353 and 354 enable the central executive to override the powers of the states to substitute temporarily a unitary system of government for the federal system. The provisions of Article 352 on external threats to national security were invoked during the 1971 Bangladesh crisis and the subsequent war with Pakistan in December of that year.[17] The references regarding internal threats to national security were invoked to justify the declaration of National Emergency that remained in effect between June 1975 and March 1977.

Emergency in the States. Under Article 356, the president, acting on the advice of the governor of a state, may proclaim a state of emergency in that state if he is satisfied that the normal process of government cannot be carried on in accordance with the constitution. This proclamation must be approved by Parliament. As a result of the proclamation, the president may assign all powers of the state to the governor, to some other executive authority, or to the central Parliament (which is in effect the central government). The new authority has the power to pass all laws for the state although these laws cease to be valid a year after the termination of the emergency. Article 356, often referred to as "President's Rule," has been invoked in various Indian states on several occasions.

Financial Emergency. Under Article 360, a state of emergency may be declared by the president, acting on the advice of the prime min-

[16] For a discussion of the emergency provisions of the constitution, see M. V. Pylee, *Constitutional Government in India* (Bombay: Asia Publishing House, 1965), pp. 636–38.

[17] See Richard L. Park and Bruce Bueno de Mesquita, *India's Political System*, 2d ed. (New York: Prentice-Hall, 1979), p. 76.

ister, if the financial or monetary stability of the country is threatened. Under this proclamation, directives may be sent to all states to resolve the financial crisis through drastic economic measures that may call for the suspension of state authority over items normally under their jurisdiction. Such directives may include salary cutbacks for all central and state government employees. A financial emergency under Article 360 has never been invoked, although drastic economic measures, such as the banning of all industrial strikes, were adopted during the Emergency between 1975 and 1977.

In addition to Articles 352–360, special ordinances may also be issued by the president on the advice of the prime minister under Article 123.[18] Such ordinances are usually promulgated when Parliament is not in session and when the president is satisfied that existing conditions call for immediate action. Presidential ordinances carry the force of legislation provided they are approved by Parliament within six weeks of its next session.

In both national security emergencies (Article 352) and emergencies in the states (Article 356), two other provisions—Articles 358 and 359—enable the president to suspend the "Fundamental Rights" of the citizen guaranteed under Article 19. Such a declaration must be approved by Parliament as soon as possible. The suspension of fundamental rights is accompanied by the right of the central government to resort to "preventive detention," that is, to arrest and detain persons to prevent them from indulging in acts—as yet not committed—that are detrimental to national security and to the maintenance of law and order. Moreover, once the emergency is declared and approved by Parliament, the constitution empowers the government to prevent citizens from appealing to the courts for the enforcement of their fundamental rights.

The issue of whether to incorporate these emergency powers into the constitution produced serious differences of opinion among members of the Constituent Assembly that drafted it.[19] Several members claimed that these provisions negated the central principles of freedom and democracy that were the basis of the Indian constitution. H. V. Kamath, one of the members of the assembly, said: "I fear that by this single chapter we are seeking to lay the foundation of a totalitarian State, a police State, a State completely opposed to all the ideals and principles that we have held aloft dur-

[18] See Robert L. Hardgrave, *India: Government and Politics in a Developing Nation* (New York: Harcourt Brace Jovanovich, 1980), p. 58.

[19] Pylee, *Constitutional Government in India*, pp. 639–55.

ing the last few decades."[20] The ability to suspend the right to appeal to the courts came under severe attack from some opponents because it eliminated a citizen's ability to gain appropriate relief or remedy for what he might consider wrongful acts committed by the government.

Other arguments against these provisions pointed out that no other similar constitution vested such extraordinary powers in the government. The Indian constitution, after all, was modeled on several western democratic constitutions.[21] The political system was modeled after the British system of government; the fundamental rights were essentially derived from the American Bill of Rights; the federal system and role of the central government were based on the Canadian system; the idea of the "preamble" that incorporated the "directive principles of social policy" was obtained from the Irish constitution; and the concept of the "finance commission" that governed the grants-in-aid to be provided by central government to the states was developed from the Australian system. None of these countries carried such far-reaching emergency provisions for national security purposes. In effect, the emergency powers written into the Indian constitution implied that a democratic constitution could be overthrown by a ruling party or leader and an authoritarian system imposed in its stead, through constitutional means.

Legislation passed over the next thirty-five years increased the emergency powers of the central government to facilitate the maintenance of national security. The first of these legal resorts may be found in the Defence of India Act, passed in 1962 immediately after the Chinese invasion in October. This act permitted the detention of any person "whom the authority suspects on grounds appearing to that authority to be reasonable, of being of hostile origin, of having acted, acting, being about to act or being likely to act in a manner prejudicial to the defence of India and civil defence, the security of the state, the public safety or interest, the maintenance of peaceful conditions in any part of India or the efficient conduct of military operations."[22]

The Defence of India Rules (DIR) derived from this act remained in force until 1968, six years after the 1962 Sino-Indian war and three years after the 1965 Indo-Pakistani war. During this period,

[20] From *Constitutional Assembly Debates*, vol. 7, p. 539, cited in Pylee, *Constitutional Government in India*, p. 639.

[21] A. H. Hanson and Janet Douglas, *India's Democracy* (New York: W. W. Norton, 1972), p. 36.

[22] Hardgrave, *India: Government and Politics*, p. 53.

these powers were occasionally used to justify "preventive detention" of persons alleged to have committed offenses unrelated to national security. During the 1971 Bangladesh crisis and Indo-Pakistani war, the DIR were reimposed, and after the Bangladesh crisis was over these rules were again used to detain persons for acts unrelated to external threats. Thus, for instance, the DIR were invoked to arrest railway workers during the nationwide railway strike organized by the All-India Railwaymen's Federation in 1974. The large-scale detentions during the 1975–1977 National Emergency were also justified under the DIR.

The 1971 war with Pakistan over the Bangladesh issue not only brought back the DIR but also saw the introduction of new legislation known as the Maintenance of Internal Security Act (MISA). The provisions of this act were used to justify the detention of several opposition leaders and to curb violence in strife-torn areas, especially in the border regions.[23] During the 1975–1977 Emergency, the powers of the central government were strengthened further through the passage of the 42nd Amendment to Article 352. This amendment suspended *habeas corpus* and made the reasons for preventive detention a state secret; a citizen could thus not appeal to the courts because he did not know the reasons for his arrest. The passage of such legislation was made possible because the Congress government under Mrs. Gandhi held a two-thirds majority (required for constitutional amendments) in the Lok Sabha after the March 1971 general elections. In a sense, such constitutional amendments were not unusual. Congress governments under Prime Minister Nehru had also carried a two-thirds majority in Parliament and had passed several amendments. However, most of the amendments under Nehru were less controversial and none affected the fundamental rights of the citizen to the same extent.

These legal instruments remained in force throughout the Emergency and were used extensively until it was lifted in February 1977. After the victory of the Janata party in the March 1977 general elections, the DIR, MISA, and the 42nd Amendment were repealed. Subsequently, however, some state governments enacted legislation that provided them with powers that were similar to the DIR and MISA. In mid-1979, Prime Minister Desai declared that such measures were necessary to deal with internal problems of law and order. The Janata government's recourse to such measures was not surprising, for it should be remembered that the Indian constitution

[23] See Hart, *Indira Gandhi's India* (Boulder, Colo.: Westview Press, 1976), p. 18.

under Article 352 continued to provide any government in office with extensive emergency powers despite the repeal of the DIR, MISA, and the 42nd Amendment.[24]

The return of Indira Gandhi to power in January 1980 did in fact bring back a reworked combination of the DIR and MISA. This was initially incorporated in the National Security Ordinance promulgated by the president in September 1980 and then embodied in the National Security Act (NSA) passed in December 1980. The NSA provided the central government with the right to detain any person in order to prevent him from acting in any manner prejudicial to the defense of India or to the internal security of the nation.[25] The act also enabled central and state governments to detain anyone in the interest of maintaining public order and the flow of supplies and services essential to the life of the community. This implied that industrial strikes and "bandhs" (efforts by political agitators to close all commercial and public activity in a city) that paralyzed the economic life of the country could be dealt with by widespread arrests and detentions.

Unlike the earlier DIR, MISA, and the 42nd Amendment, the new act contained safeguards against its misuse and provisions that attempted to ensure justice for the individual. Although members of the opposition in the Lok Sabha vigorously condemned the NSA, the new act provided individuals detained under its powers recourse to the courts if they believed their arrest unjustifiable. The new system also did not bar approach to the courts in cases involving the violation of fundamental rights.

Some of the safeguards did not last long. Following widespread terrorism by Sikh extremists from 1982 to 1984, the NSA was amended to deal with these newer and unprecedented conditions. In late June 1984, President Zail Singh on the advice of Prime Minister Indira Gandhi promulgated the National Security (Second Amendment) Ordinance.[26] The ordinance said that a detention order shall not be "deemed to be invalid or inoperative merely because one or some of the grounds is or are: (i) vague, (ii) non-existent, (iii) not relevant, (iv) not connected or not proximately connected with such persons, or (v) invalid for any other reason whatsoever."[27] The new ordinance also provided that the expiration or revocation of a

[24] Hardgrave, *India: Government and Politics*, pp 52–54.

[25] Statement reported by the Delhi Domestic Service in English and recorded in the *Foreign Broadcast Information Service* 8, no. 186 (September 23, 1980):E3.

[26] See *Statesman*, June 23, 1984; and *Indian Express*, June 23, 1984.

[27] *Hindu*, June 23, 1984.

detention order would not bar another detention order against the same person. However, the issue of the subsequent detention order would not extend beyond twelve months as in the original NSA (but twenty-four months in the "disturbed areas" of Punjab and the city of Chandigarh).

The emergency provisions in the Indian constitution and its various amendments and repeals have produced mixed results. During the Sino-Indian and Indo-Pakistani wars, these provisions enabled the government to conduct the war without internal sabotage or the breakdown of internal law and order—assuming, of course, that this was likely to have occurred. Perhaps some of the emergency measures applied to prevent sabotage during external hostilities were, in fact, over-cautious and unnecessary. These included the often arbitrary and questionable detention of Communists during the 1962 Sino-Indian war and of Muslims during the 1965 Indo-Pakistani war—not unlike the detention of Japanese-Americans in the United States during the Second World War. However, the brevity of these detentions in India and the quick release of detainees on the termination of hostilities prevented serious controversies from arising.

The application of emergency provisions for the maintenance of internal law and order has proved to be more controversial. Although it may sometimes appear dubious in motive and purpose, the imposition of President's Rule under Article 356 on more than fifty occasions in various states has enabled the central government to subdue and at times to resolve various problems of political instability and local violence in the Indian states. But in some cases the imposition of President's Rule or simply the dismissal and reconstitution of non-Congress state governments under Congress or Congress-led coalitions was allegedly engineered by the Congress government for its own purposes. Examples of this are the fall or dismissal of the Communist government in Kerala in 1959, the DMK government in Tamil Nadu in 1975 in the wake of the Emergency declaration, Farooq Abdullah's National Conference government in Kashmir in 1984, and the Telugu Desam government of N. T. Rama Rao in Andhra Pradesh in 1984.[28] In the last case, the situa-

[28] See Hanson and Douglas, *India's Democracy*, p. 122; Hardgrave, *India: Government and Politics*, pp. 89–90; S. Sahay, "A Close Look: The Exit of Farooq Abdullah," *Statesman*, July 5, 1984; and report on the fall of T. N Rama Rao's government in Andhra Pradesh in the *New York Times*, August 17 and 18, 1984. The constitutional amendment in Parliament to extend President's Rule in the Punjab was passed by a

tion clearly indicated that the Telugu Desam party had not lost its majority in the legislative assembly through defections and Prime Minister Gandhi was compelled to retract her decision to dismiss the Rama Rao government.

The suspension of the parliamentary process in these cases was confined to the regional level and elections were subsequently held within the six months to a year limit placed on the imposition of President's Rule. In the case of Assam, for instance, despite warnings of extreme violence by various Assamese political groups, elections were held in that state in February 1983, a month before the mandatory expiration of President's Rule. This decision by Prime Minister Gandhi to hold elections in accordance with the constitutional provisions led to the massacre of over 5,000 Bengali Muslims. On the other hand, the prolonged crisis and instability in the Punjab produced a constitutional amendment in Parliament to extend President's Rule to October 1985, an extra year beyond the mandatory time limit imposed by the constitution.[29] These efforts were influenced, no doubt, by memories of the disaster in Assam the previous year.

The breakdown of the political system on a national scale occurred for the first time with the declaration of the 1975 National Emergency and the suspension of the constitution. Even here the supporters of the decision have claimed that Mrs. Gandhi acted within the terms of the constitution, which provided for such powers under Article 352. Indeed, as noted earlier, the makers of the constitution had deliberately included the emergency powers under Article 352 to deal with the kind of internal violence and external aggression that they witnessed between 1947 and 1950. It was claimed by Mrs. Gandhi's Congress government that such conditions existed in 1975 and thereby justified the Emergency declaration.

Critics of the 1975 declaration have argued that this far-reaching constitutional clause permitting the suspension of constitutional rights was intended to preserve the democratic system in the long run and not to destroy it. In effect, Mrs. Gandhi violated the spirit of the Indian constitution and her actions, such as the suppression of *habeas corpus*, were far in excess of what the constitution permitted. Nevertheless, they do not deny the need for sweeping powers to be wielded by the government on occasion. For example, soon

---

vote of 326 to 69 over vehement criticism by opposition legislators (see *New York Times*, August 24, 1984).

[29] *New York Times*, August 12, 1984.

after Mrs. Gandhi lifted the Emergency and after the new Janata government repealed the DIR and the MISA that allowed for preventive detention, several states began to enact their own preventive detention laws.[30] By mid-1975, Prime Minister Desai had called for a new enactment of such laws at the center to curb spreading lawlessness.

Ultimately, after the return of Mrs. Gandhi to power in January 1980, the passage of the National Security Act at the end of that year formalized the emergency powers of government to deal with threats to internal security. But the Punjab crisis of 1984 indicated that the NSA was not sufficiently stringent to overcome legal obstacles to the detention of citizens without showing cause or to continued detention without a trial. This was demonstrated in the arrest in June 1985 of Sikh moderate and president of the Akali Dal, Sant Harchand Singh Longowal. On a petition of *habeas corpus* filed on his behalf, the Supreme Court ordered Longowal released on the grounds that the detention was not justified under the NSA.[31] As noted above, the NSA was immediately amended and Longowal was rearrested. In an unsuccessful appeal against his arrest, the defense counsel claimed that the ordinance amending the NSA violated four clauses of the constitution—Article 14 (equality before the law), Article 19 (personal freedoms), Article 21 (right to personal liberty and life), and Article 22 (certain safeguards to be provided during preventive detention).

The constitutional debate regarding emergency powers and the controversy over efforts to introduce new legislation or amendments reflect the continuing struggle between those leaders in India who are concerned primarily with the democratic process and individual freedoms and those who are concerned primarily with the maintenance of external and internal security. The issue is not whether one objective should be discarded in favor of the other, but which should take precedence. The territorial integrity and the national unity of India are important considerations, but the degree to which authoritarian measures are justifiable to achieve these ends will remain the basis of debate for some time to come.

II. The Political Debate

The political situation in India since the 1975 Emergency is unlike the period of Congress party rule during the preceding twenty-five

[30] Hardgrave, *India. Government and Politics*, p. 53.

[31] See *Hindu*, June 23 and 26, 1984.

years. In that earlier era, there was a greater measure of consensus in Parliament on the Congress government's defense policy and the conduct of external relations. There were some minor exceptions while Nehru was prime minister, when some voices of dissent were heard. For example, the Sino-Indian treaty of 1954, which embodied the five principles of peaceful coexistence but which virtually signed away the autonomy of Tibet, was criticized in Parliament by members of the Jan Sangh party and Praja Socialist Party. The PSP leader, Acharya Kripalani, for instance, had criticized the *panchshil* agreement as amounting to a "surrender of Indian rights in Tibet."[32] Similar criticisms were voiced by the same party and later by the newly formed Swatantra party following the Chinese suppression of the Tibetan revolt in 1959 and the outbreak of hostilities between India and China in October 1962.[33] There were also some criticisms by the Jan Sangh of what they perceived to be a policy of appeasement toward Pakistan by the Congress government of Nehru. Such criticisms, nevertheless, appeared token, even after the 1962 military debacle. Nehru's charismatic political appeal tended to overwhelm such political differences and a good measure of national trust continued to be placed in him.

Nehru's two successors, Prime Ministers Lal Bahadur Shastri and Indira Gandhi, enjoyed similar support in the conduct of external policy although there was comparatively greater opposition to some of their domestic policies, especially during the premiership of Mrs. Gandhi. The high point of such national support on external policies may be seen in the Indian solidarity and euphoria just before and after the 1971 military victory over Pakistan.

As we saw earlier, the declaration of the National Emergency in June 1975 by Prime Minister Gandhi was a watershed in Indian politics. Following a series of events and crises, the Indian electorate voted in the Janata party twenty-two months later—the first time since Independence that the Congress party had been displaced at the federal level. The Janata party was a hastily formed coalition of disparate parties that included the Congress faction led by Morarji Desai that had broken away from the main Congress party in 1969, the Jan Sangh, the Socialist party, the Bharatiya Lok Dal, and the Congress faction led by Jagjivan Ram that broke away from Indira Gandhi's party a few days before the 1977 elections. The Janata vic-

[32] *Lok Sabha Debates*, vol. 6, pt. 2 (May 5–21, 1954), col. 7548.

[33] *Lok Sabha Debates*, vol. 18 (August 19, 1958), col. 1676; and vols. 10–11 (December 10, 1962), col. 5134.

tory appeared to signal an end to a one-party dominant system and perhaps even the beginning of a two-party system in India. Although the Congress party had been defeated, it nevertheless obtained almost 30 percent of the seats in the Lok Sabha to the Janata party's 55 percent.

The displacement of the Congress government brought about major efforts to change the direction of Indian foreign and defense policy. As discussed above, some of the authoritarian internal security measures adopted by the previous government were reversed. This, in turn, initiated vigorous debate from the Congress opposition, now further split, in Parliament. The tenor of the debate also arose from the "federal" nature of the Janata party; it was essentially a loose amalgamation of various parties of different ideologies that had existed before the 1977 elections.[34] Indeed, this made it difficult for the Janata government to arrive at a consensus on policy. Major leaders in the party were divided on the issue of the normalization of ties with China, on relations with the United States and the Soviet Union, and on nuclear policy.

With the collapse of the Janata party in 1979 and the sweeping return to power in 1980 of Mrs. Gandhi's Congress-I party with 67 percent of the seats in the Lok Sabha, the more intense debates on security policies have remained unchanged. The possibility that another Janata-type "coalitional" party might be formed again to defeat the Congress party places greater pressure on the Congress government at the center. These varying political perspectives will be examined in terms of global, regional, and internal security.

General Defense Policy and Global Relations

The Janata government's foreign and defense policies between 1977 and 1980 tended to be a reaction to Mrs. Gandhi's internal policies just before and during the Emergency rule. Thus, the Janata government's rejection of compulsory sterilization and its restoration of democratic practices in India were soon followed by the declaration of "genuine" or "proper" nonalignment, criticism of the earlier takeover of Sikkim by the Congress government, an expressed willingness to accept the status quo along the Sino-Indian border, greater efforts to reconcile with Pakistan, and hesitant efforts to abandon India's nuclear option.

All such attempted changes in policy provoked strong protests

[34] I am grateful to Walter Andersen of the U.S. State Department for pointing this out to me.

from the Congress opposition in Parliament.[35] The opposition was strong enough in the cases of the Sino-Indian border issue and India's nuclear option to compel Prime Minister Desai to revoke his earlier statements. After Desai's declaration in Washington of his willingness to accept the line of actual control between India and China, he changed his position on his return to New Delhi amidst criticism and instead referred to the pledges made by Parliament in 1962 to recover the territories occupied by China: "I stand by the pledge for recovery and I have not given it up."[36]

Similarly, in the case of India's nuclear policy, Desai had declared on June 11, 1978, that India would not engage in nuclear testing whether for peaceful or military reasons, even if others did.[37] Under strong objections both from within his party and from Congress party members, Desai retracted his statement in July and claimed that testing and use for "peaceful purposes" were not excluded. Three days later, in yet another reversal, Desai gave assurances that nuclear testing would not take place under his government even for peaceful purposes. At the same time, he admitted that he could not bind his successors to this policy.

The Janata government also claimed a major change from Congress policy in the interpretation and application of nonalignment. In contrast to Congress declarations of "positive," "vigorous," and "realistic" nonalignment immediately after the war with China in 1962, the Janata rhetoric included such expressions as "real," "proper," and "genuine" nonalignment.[38] The Janata interpretation suggested neutrality between or equal treatment of the superpowers and a more even-handed policy regarding international technical collaboration for weapons production in India and when purchasing arms from overseas.

In seeking to redefine nonalignment in terms of neutrality, the Janata government attempted to deemphasize the security rationale

[35] See *Lok Sabha Debates*, 6th ser., vol. 13, no. 39 (April 18, 1978), cols. 304–398. An assessment of the differences between Janata and Congress defense policies may be found in Raju G. C. Thomas, "Indian Defense Policy: Continuity and Change under the Janata Government," *Pacific Affairs* 53, no. 2 (Summer 1980):223–44.

[36] *Statesman*, June 18, 1978.

[37] The fluctuating statements by Prime Minister Desai on India's nuclear intentions are reported in the *Hindu*, June 12, 1978; the *Statesman*, June 18, 1978; and the *Times of India*, July 27, 1978. For an assessment, see Ashok Kapur, "India's Nuclear Politics and Policy: Janata Party's Evolving Stance," in *Perspectives of India's Nuclear Policy*, ed. T. T Poulose (New Delhi: Young Asia Publications, 1978), pp. 170–88.

[38] For Desai's changes and clarifications of his nonalignment policy, see *New York Times*, March 25, 1977; and *Times of India*, April 8, 1977.

that lay behind the practice of this policy under both Nehru and Indira Gandhi. Nehru's formulation and interpretation of the doctrine, for instance, had rejected the concept of neutralism. Addressing the Constituent Assembly in December 1947, Nehru stated that "we have proclaimed during this past year that we will not attach ourselves to any particular group. This has nothing to do with neutrality or passivity or anything else. We are not going to join a war if we can help it and we are going to join the side which is to our interest when the time comes to make the choices."[39] According to Nehru, nonparticipation in military blocs did not also mean noninvolvement or neutrality in the cold war. Through most of the 1950s, especially under the influence of his close friend and minister of defense, V. K. Krishna Menon, Nehru's foreign policy had a pro-Communist and anti–western bloc bias.

Similarly, under the Congress government of Indira Gandhi, the earlier pro-Soviet diplomatic policy was carried further when the Soviet Union became the main supplier of arms to India. Subsequently, a treaty was signed with Moscow in August 1971 that included clauses of a quasi-military nature.[40] Despite official denials that it served any military functions, the Indo-Soviet treaty could be interpreted as an attempt to forestall possible Chinese or American military moves should war break out between India and Pakistan on the Bangladesh issue.[41] These actions were all carried out under the banner of "nonalignment," suggesting a definition of the term that ranged from neutrality to temporary and quasi-military alignment, depending on the leader and party in power and the needs of Indian security.

In contrast, the initial Janata declaration of "genuine" nonalignment suggests parallels with the Swiss or Swedish practice of neutrality during the two world wars. This type of neutrality implies military and political noninvolvement in world conflicts or, where

[39] Jawaharlal Nehru, *India's Foreign Policy: Selected Speeches, September 1946–April 1961* (New Delhi: Publications Division, Government of India, 1961), p. 24. For discussions of Nehru's nonalignment policy, see P. R. Chari, "Nonalignment and International Security," in *Role of Nonalignment in a Changing World*, proceedings of a seminar held at the India International Center, April 28–29, 1976 (New Delhi: India International Center Publications, 1976), pp. 68–81; and Raju G. C. Thomas, "Nonalignment and Indian Security: Nehru's Rationale and Legacy," *Journal of Strategic Studies* 2, no. 2 (September 1979):153–71.

[40] For the text of the treaty, see *Current Digest of the Soviet Press* 23, no. 32 (September 7, 1971):5.

[41] Maj. Gen. D. K. Palit, *The Lightning Campaign: Indo-Pakistan War, 1971* (New Delhi: Thomson Press, 1972), pp. 65–66.

some political participation is unavoidable, equal treatment of the opposing powers. In the context of the cold war, the Janata government interpreted nonalignment to mean that India either must not take sides in disputes between East and West or must treat both sides equally. Heavy military dependence on the Soviet Union and constant endorsement of its policies, coupled with criticism of the United States, would violate this concept. The interpretation and practice of Indian nonalignment has, therefore, varied under the Nehru, Indira Gandhi, and Desai administrations. Nehru's foreign policy leaned toward the Soviet Union but did not involve military dependency; Mrs. Gandhi's policy was both politically inclined toward and militarily dependent on Moscow; Desai's declared policy suggested neutrality or equidistance between East and West.

By the fall of 1979, following the collapse of the Janata government and the setting up of an interim government under the Janata faction led by Charan Singh, the Indian interpretation of nonalignment moved closer to Nehru's outlook. In a speech to the National Defence College in New Delhi, Defence Minister C. Subramaniam referred to those "who had denounced nonalignment either as immoral neutralism or fence-sitting between socialism and capitalism" and declared that Janata's practice was not different from that followed by Nehru. Denying neutrality, Subramaniam stated that Nehru's "basic concept was that India would not take sides in the super power competition and would not subscribe to the thesis of bipolarity of the international system." He then went on to claim that "a nation can be nonaligned only in other nations' rivalries and confrontations. No nation can be nonaligned in respect of a nation which threatens its security or vital national interests."[42]

The subsequent Janata redefinition of nonalignment was the result of considerable Congress criticism in Parliament. Two issues were raised in the revived domestic political debate on the meaning and application of this doctrine. First, was the Janata interpretation compatible with the prevailing strategic circumstances? Second, were the differences in definition merely political rhetoric having little to do with actual implementation of policy? The Congress opposition raised doubts about Janata policy on both counts.

During the debate on foreign policy in the Lok Sabha in April 1978, the Congress opposition leader in Parliament, C. M. Stephen,

[42] "India's Defence Strategy in the Next Decade," transcript of address by Defence Minister C. Subramaniam to the National Defence College, New Delhi, on October 29, 1979.

stated that the Janata interpretation of nonalignment was not practical.[43] It was not possible to remain "equi-distant" between the "oppressor and the oppressed," or on all issues between the United States and the Soviet Union, or on all issues in the Middle East and Africa. Similarly, Congress Member of Parliament C. K. Chandrappan claimed that Janata policy had been influenced by the efforts of President Jimmy Carter and Zbigniew Brzezinski to correct India's "tilt" toward the Soviet Union by encouraging India to be friendly toward China in keeping with American foreign policy interests.[44] Chandrappan reminded the Lok Sabha that the reason why India and China had become enemies had not fundamentally changed and he argued that the former Congress inclination toward the Soviet Union resulted from American positions on Kashmir, Goa, and Bangladesh.

Even if continued cooperation with the Soviet Union was essential from a strategic standpoint, a more pertinent question was whether the Janata government would be willing to maintain close ties with the Soviet Union in the light of Moscow's earlier support of the Emergency and criticism of its opponents—who had since come to power—as right-wing reactionaries. Initially, this problem appeared to have been resolved during the visit of the Soviet foreign minister, Andrei Gromyko, to India in April 1977, when external affairs minister, A. B. Vajpayee, said that "the bonds of friendship between our two countries are strong enough to survive the demands of divergent systems, the fate of an individual, or the fortunes of a political party."[45] Nevertheless, the new definition of nonalignment, accompanied by various diplomatic shuffles, did produce a limited move away from the Soviet Union by the Janata government.

The shift may be seen in the Janata government's tendency to obtain weapons from Western Europe rather than from the Soviet Union and in its refusal to endorse Soviet policies in Southeast Asia. The Janata government chose the Anglo-French Jaguar over the

[43] *Lok Sabha Debates*, 6th ser., vol. 13, no. 39 (April 18, 1978), cols. 315–16. See also later remarks on nonalignment by C. M. Stephen in *Lok Sabha Debates*, 6th ser., vol. 24, no. 29 (March 30, 1979), cols. 551–66.

[44] *Lok Sabha Debates*, 6th ser., vol. 13, no. 39 (April 18, 1978), cols. 305–308. See also statements by K. P. Unnikrishnan (Congress) and Moshina Kidwai (Congress), criticizing Vajpayee's foreign policy and attitude toward the superpowers in *Lok Sabha Debates*, 6th ser., vol. 24, no. 21 (March 20, 1979), cols. 275–84.

[45] See S. P. Seth, "Trends in India's Foreign Policy," *Pacific Community* 9, no. 2 (January 1978):232–48.

Soviet offer of the MiG-23 Flogger. It then signed an agreement to acquire the British VSTOL Harrier, and sought (although unsuccessfully) to purchase the American TOW antitank missiles, the West German Leopard, and the British-built Sher-i-Iran, a modified version of the rejected Chieftain tank that had been ordered in large quantities by the Shah of Iran before he was deposed.[46] In Southeast Asia, the Janata government refused to recognize the Soviet- and Vietnamese-backed regime of Heng Samrin in Kampuchea.

The return of Indira Gandhi to power in 1980 reversed some of these trends. The strategy of threatening to exercise the nuclear weapons option was restored. Also reintroduced was the older interpretation of nonalignment in which quasialignment with the Soviet Union was considered acceptable if warranted by the needs of security. The Heng Samrin regime was immediately recognized. There was a return to the previous policy of purchasing arms primarily from the Soviet Union or of manufacturing weapons in collaboration with the Soviets. The last of these reversals may be seen in moves by the new Congress government to cancel part of the Jaguar agreement with Britain and the decision to conclude a $1.6 billion arms deal with the Soviet Union in May 1980. (The arms deal, however, had been initiated by the Janata government.) These policy reversals were criticized by the parties that emerged from the breakup of the Janata party—the Janata (Socialist) party, the Lok Dal, the Bharatiya Janata Party (BJP), and the remnant Janata party. The new opposition parties, together with the breakaway Congress-U party (U for Devraj Urs, chief minister of Karnataka, who fostered the split in 1979) led by Y. B. Chavan, provided a mixed and variable external perspective in Parliament, although there continued to be a uniform anti–Indira Gandhi strain in their policy positions.[47]

[46] On overseas weapons purchases by the Janata government, see Inder Malhotra, "Planning for Defense: Changed Strategic Environment," *Times of India*, January 25, 1979; Inder Malhotra, "Anatomy of Indecision: The Striking Case of DPSA," *Times of India*, Sept. 14, 1978; and other reports in the *Times of India*, January 25, 1979; and *Hindu*, May 8, 1979. See also P. R. Chari, "Indo-Soviet Military Cooperation," *Asian Survey* 19, no. 3 (March 1979):238–39.

[47] The debate in the Lok Sabha during the Defence Ministry's budget presentation in 1983 is indicative of the opposition pressures on the government. Lively exchanges took place between Members of Parliament and the Defence Minister, R. Venkataraman, on various issues, including budgetary allocations to the navy, tank procurement for the army, U.S. arms supplies to Pakistan, the Soviet invasion of Afghanistan, and recruitment and pay scales in the armed forces. See *Lok Sabha Debates*, 7th ser., vol. 36, no. 24 (March 31, 1983), cols. 345–85, and no. 26 (April 5, 1983), cols. 350–90.

The other two opposition parties that have long had an interest in the conduct of Indian external and defense policies are the two Communist parties—the parent but now smaller and weaker Communist Party of India (CPI); and the once (but no longer) pro-Chinese faction that broke away in 1965 after the 1963 Sino-Soviet split, the Communist Party of India-Marxist (CPM). During the 1960s, the CPI essentially adopted a pro-Soviet foreign policy outlook, while the CPM generally followed a pro-Chinese line. The CPI, for instance, identified itself with the ruling Congress party in conformity with the Soviet policy of courting India. The CPM, on the other hand, although sympathetic to Beijing, maintained a judiciously ambivalent stance regarding Chinese policies in the subcontinent, for fear of being branded antinationalist. In the 1970s, the CPI continued its traditional pro-Moscow policy but the CPM began to move to a more independent outlook, due to its perception that China was becoming pro-American and less revolutionary.

In the early 1980s, the position of the Communist parties on global policy tended to be analogous to what had prevailed in the late 1960s. According to the CPI thesis, there should be no normalization of Sino-Indian relations at the expense of Indo-Soviet friendship. On the other hand, the CPM insisted that the normalization of relations with neighbors was of greater importance than the continuation of close ties with Moscow, although it had also taken a somewhat sympathetic view regarding the causes of the Soviet invasion of Afghanistan.[48] The more powerful CPM, which in the past had ruled or headed coalition governments in West Bengal, Tripura, and Kerala, joined the Janata splinter groups in opposing Indira Gandhi's authoritarian rule at home and her conduct of foreign policy, especially with respect to relations with the Soviet Union, China, and the United States. In the 1980 general elections, the CPM obtained thirty-five seats, the largest number of opposition party seats in the Lok Sabha (out of a total of 544 seats), and provided one of the more influential opposition viewpoints on foreign and defense policies. However, in the December 1984 general elections, both Communist parties lost further ground in the serious defeat suffered by all the national opposition parties.

Regional Threats and Defense Policy Responses

Just as on global political issues, domestic political perspectives on regional security issues have varied among the ruling and opposi-

[48] See *Indian Express News Service*, July 23, 1980, reported in the *Foreign Broadcast Information Service* 8, no. 151 (August 4, 1980):E4–E5.

tion parties, especially after the 1975 Emergency. Differing views are evident both with respect to the traditional Sino-Pakistani threat and with reference to newer developments in the Middle East and Southeast Asia. The near consensus of support in Parliament for the government's policy toward the traditional threats from China and Pakistan before 1975 has been displaced. The splintered Janata party, the BJP, and the CPM have urged closer ties with Islamabad. At the same time, almost all non-Communist opposition parties condemned the Congress government's policies on such issues as Afghanistan and Kampuchea.

It is significant that Indian efforts to improve relations with Pakistan and China, however, were initiated by the Congress government and not by the Janata government. This began immediately after the Indian military victory over Pakistan in December 1971 and led to the signing of the Simla Agreement in July 1972 by Prime Minister Gandhi and President Bhutto. According to the agreement, both sides would settle issues bilaterally through peaceful means and would promote greater economic and cultural cooperation. The easing of Indo-Pakistani tensions in the mid-1970s was paralleled by an improvement in Sino-Indian relations that eventually resulted in the restoration of diplomatic relations in 1976 at the ambassadorial level.

Compared to the earlier Congress government, the Janata government's policies toward these two traditional adversaries involved a faster pace of normalization rather than any major change of direction. Especially in the case of China, the quickened rate of change may have been due to the concurrent Janata efforts to reduce India's military dependence on the Soviet Union and to draw closer to the West. Meanwhile, there were criticisms and warnings by the Congress party opposition about Janata's China policy, especially after the Chinese invasion of Vietnam, which took place in March 1979 while External Affairs Minister Vajpayee was visiting China. Janata policy was alleged to be hasty and short-sighted. The leader of the Congress opposition in the Lok Sabha, C. M. Stephen, declared that the timing of the invasion and the parallels between it and the invasion of India in 1962 were a deliberate attempt to remind India of China's military superiority and predominance in Asia.[49] In keeping with the Moscow line, the CPI joined in criticizing

[49] *Lok Sabha Debates*, 6th ser., vol. 24, no. 29 (March 30, 1979), cols. 562–63. Similar criticisms of the Janata government's China policy were voiced by V. M. Sundheeran and Yadyendra Dutt (*Lok Sabha Debates*, 6th ser., vol. 24, no. 25 [March 26, 1979], cols. 371–76).

the Janata government's policy toward Pakistan and China, especially when it ran counter to Indo-Soviet ties. As opposed to Congress party doubts about China, there was less opposition to achieving better relations with Pakistan. However, criticism began to grow here as well by 1979 as reports of Pakistan's clandestine nuclear weapons programs began to be received.

The Soviet invasion of Afghanistan occurred during the change of government from Janata to Congress between October 1979 and January 1980. This did not immediately affect the trend toward better relations with Pakistan and China, but there was initial apprehension in the Indian government about U.S. proposals to send arms to these countries to stem any further advance by the Soviet Union. These fears were allayed when the Carter administration's offer of $400 million worth of arms was rejected by President Zia ul-Haq but later grew when Pakistan accepted $2.5 billion in arms from the Reagan administration in 1981.

Meanwhile, there was increasing debate in and out of Parliament over India's policies toward Pakistan and China in the light of the Soviet invasion of Afghanistan. The occupation produced a divided reaction from the opposition. Janata party leader Subramaniam Swamy called for a common Pakistan-India-China strategy to meet the new Soviet threat. Following successive visits to China and Pakistan, he suggested that the concept of a joint Indo-Pakistani defense of the subcontinent, first proposed by Field Marshal Ayub Khan to Jawaharlal Nehru in 1959, should be revived. Similarly, A. B. Vajpayee of the BJP called for the promotion of Indo-Pakistani and Sino-Indian solidarity in the face of the Soviet invasion. He saw Afghanistan's security linked with the security and independence of the countries of the region.[50] But such general proposals of the conservative BJP were not endorsed by other opposition parties such as the Janata-S(Socialist), the CPI, the CPM, or the Congress-U. Chandrajit Yadav of the Janata-S expressed fears that China and Pakistan were about to forge an alliance with the United States to the detriment of India.[51] Yadav also warned Parliament of future Chinese perfidy and demanded that China "demilitarize the Karakoram road area" linking Xinjiang with Pakistani-occupied Kashmir.

To such proposals, members of the ruling Congress-I party responded that the opposition members who called for a rapid closing

[50] *Times of India*, February 18, 1980.

[51] *Lok Sabha Debates*, 6th ser., vol. 6, no. 21 (July 7, 1980), cols. 416–19.

of ranks among India, Pakistan, and China[52] were merely pleading the American position on the regional situation. The law minister in the Congress cabinet, Shiv Shanker, declared that Indian policy was "quite clear that in the present context Pakistan cannot be armed so much as to fight the Soviet Union. Our apprehension is that whenever Pakistan is armed it leads to destabilization of the regional balance."[53]

By late 1980 and early 1981, especially after the passage of the National Security Ordinance, there was growing skepticism in Parliament about Mrs. Gandhi's claim that the danger of war loomed over the subcontinent. Most opposition members felt that external threats to India, especially from Pakistan and China, were being greatly exaggerated by the government in order to distract attention from its efforts to introduce more authoritarian measures at home. Parliamentary debates, therefore, began to focus more on the problems of internal security issues.

The Internal Security Debate

Most of the issues regarding the nature of internal security and domestic political stability have already been touched upon in the earlier sections. The Congress government's handling of insurgency and separatist movements through army actions in Nagaland, Mizoram, Tripura, and Meghalaya tended not to be controversial. To the extent that crises in Kashmir involved separatism, the Congress government's efforts to detain members of the Kashmir Plebiscite Front, the Jamiat-i-Islami, the Jamiat-i-Tulaban, or other pro-Pakistani groups again proved to be less subject to controversy. Political opposition in Parliament in such cases revolved around the efficacy of government policies rather than the need to adopt strong policies to meet such internal threats. Thus, for instance, the continued military occupation of Nagaland or Mizoram raised doubts periodically among the opposition parties about the government's ability to control the guerrilla movements in these states.

On the other hand, domestic political controversy usually stemmed from central government intervention following political instability in the states. Major political debates have occurred in Parliament over the dismissal of state governments and the imposition

[52] See report of the Rajya Sabha debates in the *Patriot*, March 15, 1980.

[53] See also statements by Eduardo Faleiro and Madhavrao Scindia, both of the Congress party, urging the government to warn Pakistan and China against escalating the Afghanistan crisis (*Lok Sabha Debates*, 7th ser., vol. 6, no. 22 [July 8, 1980], cols. 318–23).

of President's Rule. Serious dispute concerning the handling of state government crises began after the 1967 national and state elections. In the elections for the Lok Sabha, the Congress government's majority in Parliament was reduced from its usual 70 to 75 percent to 55 percent.[54] In the state elections, the Congress party failed to obtain a majority of the seats in the Vidhan Sabhas in eight states.[55] In six states opposition parties and coalitions formed governments and in others the Congress party had to forge coalitions to retain power. After several "floor crossings" by members of the legislative assembly, several state governments fell and President's Rule became frequent, a state of affairs that has continued.

The National Emergency gave further impetus to opposition groups in resisting the Congress government's dismissal of state governments. These problems, however, were not confined to Congress government rule. One of the problems faced by the Janata government upon assuming power in March 1977 was that most state governments remained under Congress rule, a situation the Janata party felt did not reflect the new mood of the Indian electorate. In particular, this became unacceptable when the election of a new president was required in 1977 after the death in office of Fakruddin Ali Ahmed, a Congress nominee. (The president of India is elected by an electoral college consisting of members of Parliament and the state assemblies.) Through presidential decrees by the acting president, several Congress governments were dissolved. Elections were subsequently held in eleven Indian states and three Union territories. State Janata parties were voted into office in the Hindi-speaking heartland, a Janata-backed CPM government displaced Congress rule in West Bengal, and regional parties were voted to power in Kashmir, Punjab, and Tamil Nadu.[56]

The return of Mrs. Gandhi's Congress party to power in 1980 produced similar efforts to topple state governments ruled by the Janata or other non-Congress parties. The problem of different parties controlling Parliament and the state assemblies was largely caused by the separation of national and state elections in 1971. Until 1967, elections to both the central and state legislative bodies were conducted simultaneously. But in 1971, Mrs. Gandhi chose to hold the national elections a year early, leaving the state elections to be held the next year, when the maximum tenure of five years for

[54] For the voting results in the Indian national elections, see Park and Bueno de Mesquita, *India's Political System*, pp. 124–28.

[55] Hardgrave, *India: Government and Politics*, p. 158.

[56] Ibid., p. 179.

both Parliament and state assemblies was due to expire. Additionally, the collapse of several state governments since 1967 because of "floor crossings" required the holding of new elections immediately thereafter. This provided the new state government a maximum life of five years from the date of the elections and again kept the next state elections separate from the national elections.

The internal security ramifications of state government instability primarily developed from the civilian violence that often ensued from the dismissal of a state government. In 1984, for example, the controversial dismissals of the National Conference government of Farooq Abdullah in Kashmir and the Telugu Desam government of N. T. Rama Rao in Andhra Pradesh produced violence in these states accompanied by the use of weapons by police and several casualties. As noted earlier, although the Kashmir government was dismissed because of its failure to maintain a majority in the legislative assembly, the dismissal of the Andhra Pradesh government had no such basis. Similarly, widespread violence in the states of Assam and Punjab from 1982 on led to the dismissal of these state governments and the declaration of President's Rule. The fall of duly elected governments in all four states and the ensuing violence generated heated controversies in Parliament on the legitimacy or excessiveness of the Congress government's policies.[57]

The deployment of army units in both Assam and Punjab also provoked opposition members to raise legal and constitutional questions.[58] On several occasions they severely criticized the use of paramilitary forces in particular in the maintenance of internal security. This was more controversial than the deployment of the army, because it suggested the growing erosion of the authority of the Indian states and the increasing tendency of the central government to move toward an authoritarian police state. Even the use of the Central Industrial Security Force generated criticism in some

[57] For debates in Parliament on these issues, see *Lok Sabha Debates*, 7th ser., vol. 26, no. 23 (March 23, 1982), cols. 435–50; *Lok Sabha Debates*, 7th ser., vol. 35, no. 12 (March 14, 1983), cols. 506–541. See also William K. Stevens, "India's Political Rage: Mrs. Gandhi and Foes Show Tempers as Nation Prepares for Fall Elections," *New York Times*, July 9, 1984; K. K. Katyal, "The Punjab Debate in Parliament: A Hardening of Attitudes," *Hindu*, July 30, 1984; S. Sahay, "A Close Look: The Exit of Farooq Abdullah," *Statesman*, July 5, 1984; and editorial in the *Statesman*, July 4, 1984.

[58] On questioning by A. B. Vajpayee of the Bharatiya Janata party, the Home Minister in the Congress government, P. C. Sethi, explained that the army was deployed in Assam under the Assam (Disturbed Areas) Act of 1955 and the Armed Forces (Special Powers) Act of 1958 (*Lok Sabha Debates*, 7th ser., vol. 35, no. 12 [March 14, 1983], col. 539).

circles in Parliament following the passage of the Central Industrial Security Force (Amendment) Bill 1983. The bill gave the CISF the same force and authority as the Border Security Force, the Central Reserve Police Force, and the Indo-Tibetan Border Force. Some Members of Parliament alleged that the new powers of the CISF were intended to break the trade union and general labor movement in India.[59] The intensity of the debate over how to achieve domestic political stability can be expected to grow in the future.

III. The Structure and Process of Security Decision Making

Apart from the nature of the political system, the basic constitutional framework, and the domestic political debate, security policies are ultimately shaped by the decision-making structure and process. The organizational framework of the decision-making apparatus and the manner in which it is used affects the contribution of various participants in the policy-making process.

The basic defense decision-making structure in India provides for interaction among various participants at three levels, usually in a hierarchical order.[60] The top or political level consists of interaction among politically elected representatives, civil servants in the bureaucracy, and military chiefs. At the middle or bureaucratic level, interaction between the relevant authorities of the civil service and the armed services takes place. The bottom or military level involves the interaction among the heads of the army, air force, and navy. The process of making and implementing Indian security policies and defense programs occurs throughout the three levels. Policy making and the allocation of resources involve interaction at all levels—political, bureaucratic, and military—but remain the ultimate prerogative of the the first level. Implementation is generally confined to the second and third levels.

The present decision-making system for defense was initially introduced at the time of Independence on the advice of both Lord

[59] See the exchange of views between the minister of state in the Ministry of Home Affairs, Nihar Ranjan Laskar, and Members of Parliament H. Hanumanthappa, Sukomal Sen, and others (*Rajya Sabha Debates*, vol. 136, no. 11 [May 10, 1983], cols. 286–308).

[60] See the assessment of former defense secretary, P.V.R. Rao, in his *Defence without Drift* (Bombay: Popular Prakashan, 1970), pp. 307–310; and his *India's Defence Policy and Organisation since Independence* (New Delhi: United Services Institution of India, 1977). A discussion of how this system worked may be found in Thomas, *Defence of India*, pp. 85–89.

Ismay, who had served as secretary to the Defence Committee of the British cabinet and was on the staff of the last British viceroy, Lord Louis Mountbatten, and Mountbatten himself.[61] Essentially, Lord Ismay recommended a three-tier committee system consisting of the Defence Committee of the Cabinet (DCC), the Defence Minister's Committee (DMC) and the Chiefs of Staff Committee (CSC). Each of these was aided by other specialized and information-gathering committees such as the Joint Intelligence Committee, Defence Science Advisory Committee, the Joint Planning Committee, and various interservice organizations to deal with pensions and personnel matters.[62] This basic framework remained for about fifteen years after Independence before it began to undergo various modifications. Before discussing each level, it should also be pointed out that decision making for internal security was undertaken separately by the Home Ministry in conjunction with the prime minister and the cabinet until the mid-1970s, when it was incorporated into a broadly based decision-making system that dealt with both external and internal security.

The Interaction of Ministers, Civil Servants, and the Military

At the top level of the decision-making system was the Defence Committee of the Cabinet. During the 1950s, the DCC did not meet regularly. Under V. K. Krishna Menon as the Minister of Defence, the Defence Minister's Committee, at the middle level, was more active in the making of policy. The development and implementation of policy by the DMC was facilitated by the fact that Menon had direct personal consultations with Prime Minister Nehru on defense and foreign policy matters. In effect, the DCC was bypassed.

The 1962 Sino-Indian war and the resignation of Krishna Menon suddenly made the DCC the most crucial decision-making organization. It was renamed the Emergency Committee of the Cabinet (ECC) during the outbreak of the 1962 war to indicate the urgency of the deliberations but also to allow for the greater participation of the chiefs of staff of the armed services in the committee. By 1976,

[61] See Lt. Gen. S. K. Sinha, *Higher Defence Organisation in India* (New Delhi: United Services Institution of India, 1980), p. 5. See also A. L. Venkateswaran, *Defence Organisation in India* (New Delhi: Publications Division, Ministry of Information and Broadcasting, 1967), pp. 80–157; and Lt. Gen. S. K. Sinha, *Of Matters Military* (New Delhi: V.S.M. Books, 1980), pp. 32–48. For the structure of operations of defense decision making, see the annual *Report· 1979–80* (New Delhi: Ministry of Defence, Government of India, 1980), pp. 4–7.

[62] See Venkateswaran, *Defence Organisation in India*, pp. 89–96.

the DCC/ECC was expanded into the much broader Political Affairs Committee of the Cabinet (PACC). This constituted a major change in security decision making at the highest level; the new committee was concerned with the entire spectrum of external and internal political and security issues. The rationale for this change was twofold. First, it was felt that external security considerations could not be separated from problems of internal political stability.[63] Often the two are directly connected—for example, when political unrest in Kashmir, Punjab, Nagaland, or Mizoram is allegedly aided or abetted by Pakistan or China. Second, it was felt that resource allocations to defense should involve an assessment of total resources available and a balanced distribution of such resources. Excessive allocation to defense that affected development or economic stability could prove disastrous for the Indian defense effort itself in the long run. The security policy-making process therefore required input from a broad range of participants.

The concept of linking the two arenas of security had even been approved earlier, by the Janata government. In October 1979, the defense minister, C. Subramaniam, stated that "the concept of national defence must be much wider than the mere protection of the country's territorial integrity and sovereignty from perceptible military threats. What needs to be shielded equally is the whole spectrum of India's political, social, economic and technological progress from pressures arising out of the play of international forces."[64] Implicit in this statement is the belief that external security and internal stability are interdependent. The PACC was expected to encompass a broad vision of all urgent and critical needs of the nation and to plan defense in this context.

Although there may have been good reasons for establishing the new committee, there were also problems associated with it. The treatment of external and internal security issues as components of one general problem has raised fears that this would eventually undermine the democratic basis of India. Various political groups have felt that blurring the distinction between the use of force to stop external aggression and the use of force to maintain internal law and order would give the ruling party or its leader considerable power. Eventually, the situation could lead to the usurpation of total power and the suspension of fundamental rights. Most of these

[63] For a discussion of this, see P. R. Chari, "Internal and External Security," *Economic Times*, October 26, 1980.

[64] Subramaniam, "India's Defence Strategy in the Next Decade," National Defence College, New Delhi, October 29, 1979.

fears have stemmed from Indira Gandhi's declaration of the Emergency.

Although the broader constitution of the committee was expected to provide more balanced economic and political planning, this advantage was thought to be offset by the possible inability of the PACC to act swiftly in case of a sudden external threat to the nation. Since all the important cabinet ministers that sat on the earlier DCC would continue to do so on the newer committee, those directly connected with defense would still be represented. However, there would be a diffusion of information on a variety of internal and external matters and a diffusion of responsibility, leading perhaps to a diffusion of expertise. With more individuals on the committee, the deliberations of the PACC would be slower and more extensive than those of the DCC. Since India has not been engaged in a new war since 1971, the ability of the PACC to deal with sudden external threats has yet to be tested. On the other hand, the PACC has had considerable experience in dealing with matters of internal stability and security, which have been the more urgent problems facing India over the last decade.

From the standpoint of defense planning, the most serious problem with the new committee is found in the reduced participation of the armed services in its deliberations. As Lt. Gen. S. K. Sinha noted, although the service chiefs are technically expected to attend the meetings of the PACC, "these occasions have inevitably got fewer because this committee is also involved with a host of other problems which are not connected with defence."[65] According to Sinha, since the institution of the new committee, the service chiefs have not been invited to attend "even when the committee is considering matters pertaining to defence." Instead, the services were often represented by the defense secretary, thus eliminating direct military input into the highest policy-making forum. As discussed below, this tends to aggravate the already strained relationship that usually prevails between the defense secretary and the service chiefs.

Finally, less well known but an important influence in the making of public policy in India at the highest level are the prime minister's personal advisors—especially the principal secretary–who form part of the prime minister's secretariat. Under Prime Minister Nehru, the principal secretary was M. O. Mathai, and under Indira Gandhi the post was held by P. C. Alexander, both men belonging to the Orthodox Syrian Christian community established on the

[65] Sinha, *Higher Defence Organisation*, pp. 6–9.

Malabar coast (now part of the state of Kerala) in the first century A.D. Following the assassination of Indira Gandhi in 1985, members of the secretariat, including P. C. Alexander, were retained by the new prime minister, Rajiv Gandhi. The importance and vulnerability of this source of policy formulation was displayed in the "spy scandal" of early 1985, when it was discovered that French agents, in conjunction with some Indian businessmen, had managed to employ certain members of the prime minister's secretariat as well as some members of the Ministry of Defence.[66] Confidential policy positions obtained from these sources on Indian weapons procurement and the Tamil separatist problem in Sri Lanka were allegedly being transmitted to both the Central Intelligence Agency and the Warsaw Pact countries. The scandal led to the resignation of Alexander and to the further tightening of security regarding the nature of the policy inputs from the prime minister's secretariat.

The Interaction of Civil Servants and the Military

At the middle level, the Defence Minister's Committee (DMC), initially proposed by Lord Ismay and instituted by the newly independent government in 1947, was intended to facilitate interaction between the Ministry of Defence and the armed services headquarters. Its functioning in the early years, especially under Defence Minister Krishna Menon, ran into difficulties. He had a general lack of respect for armed service leaders and consequently "charged happily along doing whatever he thought best."[67] The result was a marked improvement in the foundations of defense production in India (which primarily involved civilian employees) but the deterioration of civilian-military communication and coordination. The ultimate consequence of this was the military disaster in the war with China in 1962.

Subsequently, under the new defense minister, Y. B. Chavan, an informal system known as the "morning meetings" was introduced in addition to the formal meetings of the DMC. The morning meetings occurred about once a week and were attended on an ad hoc basis by the minister of defense, the minister of state for defense production, the army, air force, and naval chiefs of staff, the cabinet secretary, the scientific advisor, the additional secretary of defense, and various joint secretaries in the defense ministry. There was no

[66] For reports on the scandal, see the *New York Times*, Jan. 20 and 24, 1985.

[67] Air Chief Marshal P. C. Lal, *Some Problems of Defence* (New Delhi: United Services Institution of India, 1977), p. 24.

agenda for these meetings nor were any formal papers discussed. Although this informal setting provided an opportunity to take stock of daily developments, it generally proved to be an inadequate system to make decisions on any substantial matters.

The frequent resort to morning meetings led to a decline in the use of the DMC. After May 1974, the DMC did not meet at all and eventually this once-important committee was terminated. In its place, a new committee was formed in 1978 called the Defence Planning Committee (DPC). The DPC met under the chairmanship of the cabinet secretary and included the secretary to the prime minister; the secretaries of defense, defense production, external affairs, finance, and the planning commission; and the three chiefs of staff of the armed services. Since the DPC consists of seven high-ranking civil servants and three service chiefs, the latter feared that the preponderance of civilians would work against military input into defense policy making.[68] It was also argued that the new committee provided no direct interaction between the defense minister and the service chiefs and that the committee might eventually erode the authority of the defense minister himself, since the DPC, headed by the cabinet secretary, could report directly to the PACC.

The basic problem at the middle level of the decision-making process was the often poor relations between the civil service bureaucracy in the Ministry of Defence and the armed service headquarters.[69] The relationship has much to do with the structure of the system rather than the personalities involved, although compatible civilian and military bureaucratic heads have tended to alleviate the underlying potential for friction. To begin with, unlike counterpart systems in other parts of the world, the Ministry of Defence in India consists of civil servants and functions separately from the armed services headquarters.[70] This has not only tended to create a general lack of coordination and cooperation but has also raised suspicions and mistrust between civil servants and military officers.

Second, and more specifically, the problem lies in the ambiguous hierarchy and line of authority between the defense secretary and the service chiefs, which is also reflected in other lines of command at lower levels between the civilian bureaucracy and the military headquarters. Strictly speaking, the chiefs of staff of the army, air force, and navy are bureaucratically superior to the secretary of de-

[68] Sinha, *Higher Defence Organisation*, p. 8.

[69] See Lt. Gen. M. L. Thapan, "Question Time in War and Peace," *Statesman*, September 29, 1980.

[70] Sinha, *Higher Defence Organisation*, p. 9.

fense. However, because the defense secretary calls the service chiefs to meetings, especially in the old DMC and the morning meetings, and because he also provides the link between the service chiefs and the defense minister, in effect he has acted as the coordinator and the superior of the service chiefs. In 1984, the former vice chief of the army staff, Lt. Gen. S. K. Sinha, claimed that the corresponding hierarchical status of civil service officers versus military officers in the upper echelons was being steadily tilted in favor of the civil service. The status of secretaries at various levels in the civil service was upgraded while that of military officers was downgraded in the civil service rankings.[71]

The situation has been well described by former Air Chief Marshal P. C. Lal, who claimed that the civilians in the Defence Ministry and military officers in the service headquarters are unable to understand each other's point of view, which results in service program proposals being repeated time and again with no action taken.[72] Several other retired service chiefs and deputies interviewed have corroborated the existence of this basic problem.[73] According to Adm. S. M. Nanda, a former chief of naval staff, the relationship is such that almost all equipment needed by the services beyond maintenance and sustenance of ongoing projects requires the sanction of "some joint secretary," which thus involves service chiefs and their deputies in prolonged haggling with the Defence Ministry.[74] Consequently, as Lieutenant General Sinha has noted, a "thick layer of civil servants is interposed between the Defence Minister and the three Service headquarters, preventing direct interaction between the Minister and the Services."[75] Sinha further ob-

[71] "Ironically, whether by design or by accident, the lowering of the status of Service officers has been carried out after every war in which the Services were called upon to make the supreme sacrifice in the service of the nation. Service Chiefs became junior to Judges of the Supreme Court after the 1971 war, Lieutenant Generals were made progressively junior to secretaries in the Government of India, to Members of the Railway Board, to Chief Commissioners, and to Chief Secretaries. Similarly, Major Generals became successively junior to Chief Secretaries, Director of the Intelligence Bureau, General Manager of Railways, and to the Inspector General of Police in the States" (S. K. Sinha, "Honor the Soldier: Grievances Need to be Redressed," *Statesman*, June 25, 1984).

[72] Lal, *Some Problems of Defence*, pp. 47–48.

[73] Interviews in 1980–1981 with Admirals S. M. Nanda and A. K. Chatterji, Air Chief Marshal Arjan Singh, and Lt. Gen. A. M. Vohra. See also Gen. J. N. Chaudhuri, *India's Problems of National Security in the Seventies* (New Delhi: United Services Institution of India, 1973), pp. 44–48.

[74] Interview with Adm. S. M. Nanda, December 1980.

[75] Sinha, *Higher Defence Organisation*, p. 10. This inability of the service chiefs to

serves that although the service chiefs could conceivably approach the defense minister directly, "the fact remains that all papers referred to the Ministry by the Service Chiefs have to be processed by civil servants headed by the Defence Secretary." According to some military officers, this situation has arisen largely because of the absence in the Indian system of a "chief of defense staff" who could coordinate the activities of the three service headquarters and provide the necessary hierarchical status, political clout, and link between the military and the civilian-political wings of the defense decision-making system.

Civil service officers in the Defence Ministry claim that such assessments are exaggerated.[76] Their response to the problems expressed by the armed service headquarters is probably best stated by K. Subrahmanyam, a former secretary for defense production.[77] He suggests that the present situation derives from the basic structure of the system of government in India and particularly from parliamentary procedures in the allocation of resources—stuctures and procedures that apply to all ministries and departments. For example, in contrast with the United States, where defense expenditure is based on "obligational authority," in India approval is based on the "flow of expenditure" and not on total expenditure. According to Subrahmanyam, when the U.S. Congress approves a plan for the MX missile system, for example, financial provision would be made for the total negotiated amount needed to develop and deploy the system. On the other hand, in India if a decision is made to

---

make what they consider substantial inputs into defense programs, especially on a day-to-day basis, is best illustrated by the story narrated by Brig. J. P. Dalvi: "I know of one case in which a relatively junior civil servant had the audacity and impertinence to start his note by saying that 'I do not entirely agree with the proposals of the Chiefs of Staff,' and then went on to suggest a modification of the proposal. The Minister agreed with the civil servant but the Chiefs were never told why the Minister's final orders differed from their original recommendation They never saw the relevant file to find out who was the genius who knew better than they" (Brig. J. P. Dalvi, *Himalayan Blunder* [New Delhi: Hind Pocket Books and Orient Paperbacks, n.d.], p. 357; also in Sinha, *Higher Defence Organisation*, p. 10). For more details of the frustrations experienced by the Indian army with the Ministry of Defence, see Dalvi, pp. 353–60.

[76] This view was also expressed to me in 1984 by P. R. Chari of the Indian Administrative Service, who had been a joint secretary and director in the Ministry of Defence in the early 1970s.

[77] Interview with K. Subrahmanyam, January 1981. For further details of the constitutional aspects and workings of this system, see report of the former financial advisor in the Ministry of Defence, D. S. Nakra, *Defence Budgeting in India* (New Delhi: United Services Institution of India, 1979), pp. 2–5.

purchase three hundred tanks, Parliament would sanction only the amount needed for the current fiscal year. If fifty tanks are scheduled to be purchased in the current year, the cost of this number is all that is approved. Subsequently, new amounts must be sanctioned yearly until all three hunded tanks are eventually purchased.

Therefore, in the present Indian system, once allocations are made to the services for various expenses, substantial deviations from the programs proposed or extra amounts demanded beyond that approved for the year must be approved by the relevant higher political authorities, whatever the comparative rankings of civil service and military service officers. According to Subrahmanyam's estimates in 1981, anything in excess of Rs. 5 crores ($6.25 million at the 1981 exchange rate of $1 = Rs. 8) must be approved by the cabinet; anything between Rs. 2 crores ($2.5 million) and Rs. 5 crores must be approved by the finance minister; anything between Rs. 1 crore and Rs. 2 crores must be approved by the Defence Ministry; and anything less than Rs. 1 crore must be approved by the financial advisor in the Defence Ministry. Therefore, according to Subrahmanyam, the civil servants in the Defence Ministry are merely fulfilling their obligations prescribed by the system. In effect, the system allows the Finance Ministry to control the Defence Ministry, and the Defence Ministry to control the armed services headquarters—all through the power of resource allocation.

Three proposals have thus far been made to alleviate this condition. The first proposal, as expressed by Air Chief Marshal Lal on the suggestions of other service officers, is to allow some military officers, perhaps even retired ones, to take over civilian jobs in the Defence Ministry. The problem with this, according to Lal, is that any large-scale shift of military officers to civilian functions could impair the judgments of the former in both their civilian and military functions: "The fact of the matter is that civilian officers in the Ministry of Defence and other departments of government perform as specialized a function as military officers in their units and formations. Unlike men in uniform, they can and do remain in particular departments for many years at a time."[78] Lal's views, however, are not generally endorsed by other service officers, who feel there should be greater integrating and overlapping functions between the Defence Ministry and the service headquarters.

A second suggestion sometimes considered is that the level of discretionary funds for each of the three services be substantially in-

[78] Lal, *Some Problems of Defence*, pp. 46–47.

creased and that a financial advisor from the Ministry of Finance be assigned to each of the three armed services headquarters to advise and oversee all their expenditures.[79] This would be in contrast to the present arrangement, in which only one financial advisor from the Ministry of Finance is assigned to the Ministry of Defence and not to the service headquarters. Greater discretionary funds for the services would reduce the constant contact and friction between the civilian and military branches.

According to a former financial advisor in the Defence Ministry, D. S. Nakra, the present system is unworkable because the services and the Defence Ministry make financial demands in Parliament on the basis of estimates. They are then forced to work within their statistical projections, which often will have been trimmed earlier by the Finance Ministry. Nakra cited one case in which the army had asked for Rs. 80 crores including a foreign exchange component of Rs. 40 crores. A compromise was reached of Rs. 20 crores including a foreign exchange component of Rs. 12 crores. "Before even the first phase was completed, an emergency exposed its unsoundness and it was dropped, and sanction was given for a much larger amount with a much higher foreign exchange component." Therefore, according to General Sinha, the "overcentralization of financial control" tends to be time-consuming and wasteful.[80] The prospect, however, of changing the present arrangement to give the services greater financial autonomy is still remote. The reason stems primarily from a somewhat excessive desire on the part of politicians and civil servants to maintain a tight control over the military.

A third proposal put forward, especially by the army, is the appointment of a chief of defense staff with his own secretariat, who would provide a combined-service outlook and direct line of communication with the Defence Ministry and the PACC. This would enable the Defence Ministry and the armed services headquarters to deal with each other on an equal basis. This idea, as will be seen in the discussion that follows, has problems of its own.

The Interaction within the Military

The third level of the decision-making process involves the interaction of the three armed services headquarters in the Chiefs of Staff Committee (CSC). Again, the origin of this committee goes

[79] See Sinha, *Higher Defence Organisation*, pp. 23–25, and D. S. Nakra, "Fnancial Management in the Defense Services," *United Services Institution of India Journal* (July-Sept., 1973).

[80] Sinha, *Higher Defence Organisation*, p. 24.

back to the recommendations of Lord Ismay in 1947. Initially, each service had a separate commander in chief and the three chiefs were given equal status, although in ranks they varied from a four-star general for the army to a three-star air vice marshal for the air force and a three-star vice admiral for the navy.[81] For various reasons, especially because the term "commander in chief" reminded the Congress government of the tremendous powers wielded by the British commander in chief before Independence, the designation was changed to the chief of staff.

The chiefs of staff are concurrently principal military advisors to the defense minister as well as the commanders of each of the three services.[82] In theory, this is a suitable arrangement because the chiefs of staff both propose programs for their services to the defense minister and are responsible for carrying them out. According to Air Marshal Lal, "the fact that the Chiefs are accountable for the action taken on the basis of their own advice and plans, makes for what may be called 'responsible planning,' a salutory practice that can do much to bring a sense of realism to military thinking."[83] The CSC is advised by other committees, such as the Joint Planning Committee, the Joint Training Committee and at one time also the Joint Intelligence Committee, which after the 1962 Sino-Indian war was placed directly under the Defence Committee of the Cabinet.

Although the arrangement generally provided for the satisfactory daily administration of operations in times of peace, there were visible drawbacks in times of war, especially during the 1965 Indo-Pakistani war. According to General Sinha, "the 1965 war virtually saw the Army and the Air Force fighting two separate wars and the Navy only marginally involved."[84] The air force was more concerned with establishing air superiority by attacking Peshawar and Sargodha air bases in Pakistan than with providing critical ground support for beleaguered army divisions in Jammu and Kashmir.

A similar lack of coordination, although with a successful outcome, occurred during the 1971 Indo-Pakistani war. In the Rajasthan-Sind sector, the army launched a thrust across the desert toward Karachi, the navy raided Karachi harbor, and the air force destroyed the oil storage depots of Karachi—each operation conducted independently of the other. However, General Sinha felt that the overall success of the Indian armed services in the 1971 war

[81] Ibid., p. 6; Venkateswaran, *Defence Organisation in India*, pp. 138–40.

[82] See Lal, *Some Problems of Defence*, pp. 8–20.

[83] Ibid.

[84] Sinha, *Higher Defence Organisation*, p. 19.

had more to do with fortuitous circumstances, conditions that may not be repeated in the future. Although Air Marshal Lal subsequently claimed that the 1965 problem was largely created by General Chaudhuri's tendency to take action during the war without consulting with the other two services, these occurrences nevertheless had much to do with the separation of the three service headquarters, coordinated only by the CSC—consisting of three equal chiefs chaired on a rotational basis by the most senior chief of staff.

Under these circumstances, the solution often proposed has been the controversial one of setting up a chief of defense staff system (CDS). This has been enthusiastically supported by the army, but has received mixed reactions in the air force, navy and the defense ministry. The proposal has also generated some apprehension at the political level about potential military takeovers.

The army's case for a CDS system rests on two considerations: the need to have one center for military command, control, and coordination that would integrate the strategies of the three armed services; and the need to project a common and combined military posture to the political and civil service authorities instead of the present independent and sometimes disparate projections by the three services.

The former chief of army staff, Gen. J. N. Chaudhuri, who commanded operations during the 1965 Indo-Pakistani war, claimed that "the greatest fault" of the present system was that the service chiefs were expected to combine two roles in one.[85] This dual role was workable in the 1950s when the services and their anticipated military operations were much smaller; under present conditions one man should not be expected to fulfill both functions. Chaudhuri also pointed out that "very often the functions can be incompatible, for an honest Chief of Staff can disagree with himself as an honest Commander in Chief." Chaudhuri, therefore, suggested that a CDS would help to separate the two functions. The CDS would provide both the needed coordination among the armed services and the link between the combined military and the political-civilian authorities and the service chiefs could "concentrate wholly on operations, logistics, and alternative planning for the future."[86] Similarly, General Sinha writes that experiences in Britain, the United States, and elsewhere point to the fact that "defence policy consists

[85] Chaudhuri, *India's Problems of National Security*, p. 49.

[86] Chaudhuri, *India's Problems of National Security*, p. 50. See also views of Maj. Gen. S. N. Antia, "National Defence: A Viewpoint-I," *Indian Express*, June 27, 1980.

of a great deal more than the pooled views and agreed compromises of the three Services."[87] He adds that under various designations, of which the most common is chief of defense staff, "all other countries of the world—democratic, socialist or developing—have introduced a professional coordinator or head, by having a top brass for the Services."[88]

Arguments against establishing a CDS system have come primarily from politicians and the other two armed services, the air force and the navy. The political and civilian arguments against the CDS system stem first from the fear that the creation of such a system could eventually undermine the democratic system. One powerful figure would command the entire military and would therefore be in a position to coordinate a military takeover of the government. This opposition arises from historical experience under the British raj. Under the British, the commander in chief of the Indian army, the only major military force in India before the Second World War, was the most powerful political figure in India after the British viceroy. The situation led to a major struggle for power in India at the turn of the century between Viceroy Lord Curzon and Commander-in-Chief Lord Kitchener. Eventually Curzon asserted his authority, but the episode demonstrated the threat that a powerful military commander could hold over the political head of the state. From a political standpoint, it was clear that the British colonial arrangement would not be acceptable in an independent and democratic India. A second argument against the CDS predicted that the elimination of the present chiefs of staff system with its dual function of both proposing programs and being responsible for their implementation would produce less pragmatic proposals, since the CDS would not be responsible for carrying out what he had proposed.

The military argument against the creation of a CDS sys*em was also twofold. First, as Air Chief Marshal Arjan Singh has pointed out, such a system may be necessary only for a global power, which may need to conduct combined land, air, and sea military operations in distant strategic theaters requiring a common command control center at home.[89] Second, as both Air Marshal Lal and Admiral Nanda have commented, since the army is much larger than the air force and navy combined, a CDS would always be an army

[87] Sinha, *Higher Defence Organisation*, p. 15. Lt. Gen. A. M. Vohra expressed an identical view during an interview in December 1980.

[88] Sinha, *Higher Defence Organisation*, p. 16.

[89] Interview with Air Marshal Arjan Singh, December 1980.

man or only nominally connected with the other two services.[90] (In the 1970s, the army constituted almost 90 percent of total military manpower and received about 65 percent of the annual defense budget on combined revenue and capital accounts.) Military strategy would then invariably be biased in favor of army perspectives, manpower, and resources. This could prove detrimental for devising balanced and effective military strategies and could lead to the neglect of the other two services.[91]

The army, in turn, has responded to these arguments and is supported also by some air force and naval chiefs, such as Adm. A. K. Chatterji. In the first place, as General Sinha has argued, given the overwhelming size of the army in India, a military takeover—if it could be engineered—would not need to take into account the other two services. Sinha has claimed that it "is hardly realistic to suggest that a Chief of the Defence Staff can be a greater threat to democratic institutions than a Chief of the Army Staff."[92] Second, the advantage of locating the functions of policy formulation and program implementation in the same authority was claimed to be spurious. Sinha has pointed out that the basic structure of the government of India involves a separation of these two functions with "advice being the prerogative of various ministries, and execution the responsibility of the departments attached to these ministries."[93] In any case, such problems, according to the army, could easily be corrected by providing the chiefs of staff of the three services direct access to the defense minister whenever they felt that programs proposed by the CDS were unrealistic or inadequate.

On the irrelevance of a CDS system for a regional power, Generals Chaudhuri, Sinha, Thapan, Vohra, and others have contended that unified command, control, and coordination are just as important for regional military operations as for distant theaters of war. Finally, regarding the fears of the other two services that a CDS would lead to further army dominance, the supporters of the CDS have proposed that the position could be established on a rotational basis so that each of the service chiefs would in turn inherit the CDS

[90] Interviews with Admirals S. M. Nanda and Air Chief Marshal P. C. Lal, December-January, 1980–1981.

[91] A cohesive and comprehensive argument against the CDS is provided by Lal in *Some Problems of Defence*, pp. 39–45.

[92] Sinha, *Higher Defence Organisation*, p. 19. Additionally, according to Sinha, India's size and complexity precludes the possibility of military coups. See also Sinha, *Of Matters Military*, pp. 29–31.

[93] Sinha, *Higher Defence Organisation*, p. 21.

position. Moreover, Sinha has insisted that "it is an insult to the maturity and vision of our senior officers of the three Services to even remotely suggest that anyone of them assuming the top appointment would not do justice to the other two Services."[94]

A Proposed National Security Board

It should be apparent that the defense decision-making system in India, at all three levels of the hierarchy, has drawbacks and generates political differences among the participants. Of the levels' three major issues (i.e., controversy over the mandate and representation in the PACC, the strained relations between the civil service and the armed services, and the proposal for a CDS), it is the last that is probably the most easily resolved. Air force and naval fears of the creation of the CDS could well be allayed by the rotational system proposed by the army. Meanwhile, the growth of the services and the increasing magnitude of defense operations may soon make a CDS system inevitable. Relations between the civil service and the armed services appear less amenable to a short-term solution in that a major reorganization of the Defence Ministry itself is called for. The armed services headquarters would have to be integrated with the Defence Ministry, with civilians and military officers fulfilling various related roles side by side, a change that could not happen overnight but is not beyond the realm of possibility. At the highest level, on the other hand, the Political Affairs Committee of the Cabinet appears to be firmly established. The PACC appears to be adequate, at least in dealing with internal security issues, which have been the main problems faced by India since 1975. A return to the more narrowly conceived Defence Committee of the Cabinet is only likely to occur if the PACC proved to be ineffective during a major war or if a substantial change of government or governmental thinking occurs.

The ultimate solution to problems at all three levels may be the constitution of a National Security Board, a system that has been sometimes considered in recent years.[95] However, even here the composition of such a body and its agenda would be subject to controversy. Proposals range from a body that would be not unlike the PACC but would include only the prime minister, the minister of defense, the minister of external affairs, and the minister of home affairs and would involve greater participation by the service chiefs to

[94] Ibid., p. 20.

[95] See the *National Herald* (Lucknow), September 24, 1980.

a more elitist body that would be headed by the prime minister and consist mainly of defense specialists, both civilian and military. A National Security Board of the latter type would provide defense specialization, the coordination of varying viewpoints of both civilian and military organizations, and also the necessary bridges between the military, the civil service, and the highest political authorities. Proponents of the National Security Board have argued that it would provide functions and advantages that the PACC, the DPC, and CDS do not currently offer while also avoiding some of their failings.

CHAPTER FIVE

Military Perspectives and Defense Programs

As DISCUSSED in Chapter Two, the shift in Indian security perspectives from a strategic environment that centered around the Sino-Pakistani threat to a more extended spectrum that included military developments in the Middle East, Central and Southeast Asia, and the Indian Ocean implied the need to revise India's defense plans and programs drastically. The transformation of the strategic environment has consequently produced various proposals from the Indian armed services for revisions in their roles, force levels, and weapon systems. The new environment has raised the issue of whether or not to play out India's nuclear option; any assessment of current and future Indian defense programs must also take into account the programs in the Departments of Space and Atomic Energy.

Political maneuvering among the military bureaucracies for funding for proposed programs calls attention to the defense budgetary allocation and distribution process. The military services must first convince the Defence Ministry of the viability of their proposals; the Ministry of Defence must then convince the cabinet and especially the Finance Ministry of the necessity of the overall defense budget; and finally, the entire defense budget must be submitted to Parliament for debate and approval. It is in the parliamentary forum that projected strategic roles of the services are translated into monetary allocations and specific defense programs.

1. Strategic Perspectives and Defense Roles

The present roles of the army, air force, and navy continue to be influenced by British legacies, although to a much lesser degree than in the 1960s. Ironically, the present strategic picture resembles that faced by Great Britain at the turn of the century. In their efforts to preserve the Indian Empire, the British sought to prevent a Russian advance through Afghanistan or Persia, the historical entry points for centuries of invaders. Similarly, the British encouraged the autonomy of Tibet to create a buffer between India and the weak Chinese Empire. Threats by sea were secured by the powerful Royal Navy, which commanded the Indian Ocean.

These objectives and policies were pursued by the British through a large standing Indian army, commanded for the most part by British officers.[1] The Indian army was used effectively not only to guard the land frontiers of India but more widely to serve the far-flung British Empire overseas. The development of the Royal Indian Air Force (RIAF) and Royal Indian Navy (RIN) was much less spectacular. Both services were neglected through much of British rule in India. The RIAF received some attention only after the Second World War began, when the air force was expanded from one squadron to ten throughout India. However, the British saw little reason to expand the RIN because the Royal Navy already successfully controlled the Indian Ocean.

There were also reasons related to internal security for the British policy toward the development of the Indian armed services. The Indian National Mutiny in 1857 had for the first time united various religious and linguistic groups in India, and the British thus found it expedient to expand primarily the Indian army, which could more easily be broken up into various regiments consisting exclusively of certain caste, religious, and linguistic groups. "Divide and rule" was a policy that could not be readily extended to the RIAF and the RIN. For instance, unlike army regiments that could be divided along social lines for whatever excuse cited by the British (some of which made sense, such as different eating habits), it would have been difficult to have a frigate full of Sikhs, a destroyer full of Punjabi Muslims, bombers with all-Maratha crews, or submarines manned only by Gurkhas. Since the RIAF and the RIN tended to be unified, to promote their growth might have threatened the empire itself.

In retrospect, the rationale underlying the composition of the Indian and Pakistani armed forces out of the British Indian armed forces after Independence would seem dubious—except for the justification that they each inherited similar and proportionate military organizations and each now constituted a military as well as a political threat to the other. The threats to both states were no longer seen to stem primarily from the Soviet Union and China across the Hindu Kush and the Himalayas and at least in the Indian case, the military posed no internal threat to civilian supremacy. The Indian

[1] See Raju G. C. Thomas, "The Armed Services and the Indian Defense Budget," *Asian Survey*, 20, no. 3 (March 1980):280–97.

and Pakistani armies now faced toward each other along the Kashmir cease-fire line, the divided Punjab plains, and the desert border between Rajasthan and Sind. Overlooked was the vulnerability of East Pakistan and the southern coastal flank of the Indian peninsula that seemed to call for the expansion of the navies of both countries. Also overlooked was the fact that despite three thousand years of invasions via the Khyber Pass, from the early Aryans through the Greeks, Afghans, Persians, and Turks, the most recent invasion of the subcontinent—that of the Europeans—had come by sea. And inadequately considered was the fact that the strategic role of air power was relatively new and needed to be reevaluated after the departure of the British.

The colonial legacy to the post-Independence Indian armed services lingers, though there have been moderate changes in strategic roles and force postures, especially since the 1962 Sino-Indian war. The trends in defense strategies and the development of each service's role over the last three decades are discussed further below. Of particular concern is whether present conditions call for substantial revisions. Also relevant to this discussion is the growth of Indian nuclear weapons capability and its effects on conventional defense strategy.

Strategy and Role of the Army

Following Independence and partition, the Indian army first saw military operation in Kashmir during the winter of 1947–1948. In many ways, this was the army's first test under conditions different from those under British rule. Under the British, the Indian army's principal campaigns were to defend the frontiers of the subcontinent; this was mainly in the northwest against Afghan encroachments that had continued for over a century and later in the northeast against the Japanese during the Second World War. In addition to frontier defense, the army saw extensive military action during the two world wars in Europe, North Africa, the Middle East, and Southeast Asia.

In contrast, the first Indo-Pakistani war was within the subcontinent and resembled a civil war. The British Indian army at that time was still in the process of being divided and distributed between the two newly independent states. The Muslim officers, approximately 33 percent, were offered a choice of remaining in India or opting for Pakistan.[2] Almost all of them chose to join the new Pakistan

[2] Figures on the distribution of military forces and equipment between India and Pakistan after the partition of British India are from Lorne J. Kavic, *India's Quest for*

army. The remaining Hindu and Sikh officers remained with the Indian army. Similarly, troops in predominantly Muslim regiments were sent to Pakistan; Hindu (including Gurkha) and Sikh regiments remained in India, with the exception that some Gurkha regiments continued to serve the British in Britain and in their other overseas territories. Meanwhile, senior British officers retained command of the Pakistani and Indian armies until the early 1950s. Despite the confusion on both sides in the hasty division, sufficient units of the post-Independence Indian army were quickly deployed in Kashmir to hold off the invasion by the Pakistani army and other armed Azad Kashmiri "liberators." A cease-fire was eventually called in 1948 along what is more or less the present line of control. Throughout these operations, the Indian army acted primarily on its own with some support from the Indian air force.

The role of the army—as indeed of all three of the Indian armed services—was subsequently predicated on the experience of this first war. The only major threat to India was perceived to come from Pakistan, and since the composition and proportions of the Pakistani armed services were similar to the Indian armed services, future Indo-Pakistani wars were expected to be primarily based on land, engaged and directed by the army. Even after Pakistani acquisition of several squadrons of American aircraft for its air force and counterpurchases by the Indian air force, the importance of land operations and consequently of the army's role continued to be emphasized in the defense of India.[3]

That the army dictated military strategy in India became evident in both the 1962 Sino-Indian war and the 1965 Indo-Pakistani war. The assumption—albeit principally a political rather than a military one—that the only major threat to India came from West Pakistan found the Army with only a small force in eastern India, incapable of defending the few mountain passes agaist the rapid Chinese advance in 1962. Even after reinforcements were flown in, the army lacked the proper equipment and even appropriate clothing to fight at high elevations along the Himalayan ranges. Fortunately, in the Ladakh area of Kashmir, the army was better prepared and

---

Security: Defense Policies, 1945–1965 (Berkeley and Los Angeles: University of California Press, 1967).

[3] According to Air Chief Marshal P. C. Lal, India's military thinking in the 1950s "was conditioned by the conviction that any threat that might arise to our security would be overland, and that too in the J & K–Punjab region. The Army was seen as the principal protagonist with the Air Force in a supporting role" (Lal, *Some Problems of Defence* [New Delhi: United Services Institution of India, 1977], p. 57).

equipped and was able to hold back the Chinese invasion. There were no air operations by the air force in the Sino-Indian war except for transport and supply-dropping missions. The performance and the failure of 1962 was almost exclusively an army affair.

During the 1965 Pakistani war, the military operations were more extensive and this time the air force was also involved. Although the Rann of Kutch military skirmish between India and Pakistan in April of 1965 took the Indian army by surprise, its reaction was fast and effective when major hostilities broke out in September. However, according to P. C. Lal, who was then air vice chief marshal under Air Chief Marshal Arjan Singh, the army again planned most of the operations. As during the 1962 war, the army under the chief of army staff, Gen. J. N. Chaudhuri, continued to believe that land wars were its exclusive concern and that the role of the air force was subordinate to the needs of the army. General Chaudhuri's decision even to bring in the air force in support of the ground operations was belated. Only when the Pakistani army, equipped with American Patton tanks against Indian deployments of French AMX-13 tanks, threatened to cut the Jammu-Srinagar highway in the Chamb-Jaurian sector did the Indian army call in the air force for ground support. With the Indian decision to extend the war beyond Jammu and Kashmir, army reinforcements were ordered in from bases in central and southern India; but these operations in Rajasthan and Punjab were mainly an army affair with some air force support.[4]

This lack of army–air force coordination was corrected to some degree by the time of the 1971 Indo-Pakistani war. According to Lieutenant General Sinha, the 1971 situation had much to do with the presence of compatible personalities among the chiefs of staff (Gen. Sam Manekshaw, Air Marshal P. C. Lal, and Adm. S. M. Nanda) and the ability of the service chiefs to get along well with the prime minister, defense minister, and defense secretary (Indira Gandhi, Jagjivan Ram, and K. B. Lall). Here too operations tended to conceal inadequate coordination even where it was successful, as in the Sind sector, a case mentioned above.

Army operations in East Pakistan were particularly successful because of several factors. The Indian army outnumbered the Pakistani army, during the first few days of the war the Indian air force was able to decimate the small Pakistan air force stationed there,

[4] Lt. Gen. S. K. Sinha, *Higher Defence Organisation in India* (New Delhi: United Services Institution of India, 1980), p. 19.

and the navy attacked and bottled up entrances to East Pakistani ports, preventing reinforcements. These advantages, of course, do not deny the ingenuity of Lt. Gen. Jagjit Singh Arora's land strategy. The three Pakistani divisions stationed in the east were drawn out and strung along the borders before the outbreak of the war. General Arora's invading forces subsequently encircled the Pakistani forces pocketed along the borders without attempting to draw their fire. Meanwhile, the main columns of Indian forces drove directly to Dacca from all directions.[5] The siege of Dacca brought about a quick surrender from Pakistan's Lieutenant General Niazi, who saw the prospect of relief by outside forces as hopeless. Throughout these operations in East Pakistan, the Indian army was greatly aided by the Bangladesh guerrilla forces known as the Mukti Bahini.

In the Sino-Indian war of October-December 1962, the Indian army had fought alone and suffered defeat. There was subsequently a decade of steady rearmament of all three services in varying degrees, until the third Indo-Pakistani war broke out in December 1971. In this war the army was not only victorious but for the first time saw action in combination with the other two services. The present and future role of the army, therefore, mainly concerns newer divisions of resources and responsibilities to the air force and navy, and the type and level of command and coordination that is desirable among all three armed services. These responsibilities and capabilities need to be assessed against both the traditional threats to India and the evolving extended strategic circumstances. Regarding the traditional perspective on threats to India, the Army's capabilities until 1981 may be considered to be comparatively stronger than at any earlier time. This situation began to erode with the Reagan administration's decision to supply Pakistan with M-48A tanks, armored vehicles, and other antitank weapons and to transfer an unspecified amount of non-lethal military equipment and technology to China.

Except for these more recent events, the Indian army has appeared to have both qualitative and quantitative superiority over the Pakistani army and the ability to stall Chinese ground forces along the Himalayan heights effectively. More optimistic assessments be-

[5] For the Indian army assessment of the military operations in East Pakistan, see Maj. Gen. D. K. Palit, *The Lightning Campaign· Indo-Pakistan War, 1971* (New Delhi: Thompson Press, 1972); Maj. Gen. Lachman Singh, *Indian Sword Strikes in East Pakistan* (New Delhi: Vikas Publishing House, 1979); and Maj. Gen. Sukhwant Singh, *The Liberation of Bangladesh* (New Delhi: Vikas Publishing House, 1980).

fore 1981 even suggested that the army could not only defeat the Pakistan army in the west but also could make incisive thrusts into Tibet across the Chinese borders in the north.[6] Although there have been no major increases in the number of armed personnel in the army, these views are based on improvements in the army's artillery and armor and on corresponding deterioration in the quality of armor on the Pakistani and Chinese sides.[7] Especially along the Chinese front, improvements in communications and supply lines on the Indian side versus no similar advance on the Chinese side, together with Beijing's preoccupation with the Soviet threat and problems with Vietnam and Southeast Asia, would tend to place the Indian army in an advantageous position against China's ground forces in any future Sino-Indian war.

There are also more pessimistic assessments about the army's ability to meet its traditional threats. In the first place, the pessimists point to both the increase in the size of the Pakistan army and to the concentration of Pakistani forces within a now more restricted territorial border. No longer does Pakistan have to worry about the defense of its former province of East Pakistan, where in the past one to three divisions had been located. On the other hand, Bangladesh has raised five divisions and there were tentative plans to raise yet another five in the next few years. On the Chinese front, the situation according to these assessments has not fundamentally changed. There has been no increase in the number of Indian mountain divisions facing Chinese forces.

These conditions have prompted some military commentators, such as Lt. Gen. M. L. Chibber, to come up with an approximate order of battle of ground forces in and around the subcontinent as of 1979. The more recent military balance on the subcontinent has approximately the same proportionate distribution. See Table 5.1.

To criticisms that the sizes of the divisions differ from country to country and that the quality of equipment may vary, the more cautious assessments of the traditional threats point out that variations in size apply only in the Chinese case. But even here memories of the Indian military defeat in 1962 raise fears that the organization and quality of Chinese divisions may be superior to Indian divisions. In contrast, the quality of armor in the Indian army (the basic In-

[6] Interviews with Ravi Rikhye and Sreedhar Rao at the Institute for Defence Studies and Analyses, December 1980.

[7] For an assessment of the Indian government's decision to restrict the size of the Indian army, see Lt. Gen. M. L. Thapan, *The Imposition of a Manpower Ceiling on the Army* (New Delhi: United Services Institution of India, 1976).

TABLE 5.1
Ground Forces in and around the Subcontinent

PAKISTAN[a]	13 2	infantry divisions armored divisions (3 more infantry divisions face Afghanistan)	INDIA[a]	15 2	infantry divisions[b] armored divisions
CHINA[a]	26	infantry divisions in the south and southwest region (probably about 10 divisions directly facing Indian divisions)	INDIA	10	mountain (infantry) divisions
BANGLADESH	5	infantry divisions	INDIA	4	infantry divisions (mainly deployed in Assam, Nagaland, Mizoram, etc.)

SOURCE Lt Gen. M.L. Chibber, *Paramilitary Forces* (New Delhi: United Services Institution of India, 1979), pp. 2–4. Chibber's assessment was derived from *The Military Balance* (London: International Institute for Strategic Studies), I have updated and revised this analysis based on *The Military Balance: 1985–86*.

[a] Indian and Pakistani infantry divisions consist of about 17,500 men; the armored divisions consist of about 15,000 men and about 200 tanks A Chinese division consists of about 12,700 men, and an armored division of about 9,200 men and 270 tanks

[b] Indian infantry divisions directed against Pakistan are normally stationed in central and southern India during peacetime and moved up to the front during wartime.

dian-made Vijayanta tank and the Soviet-supplied T-54/55 and PT-76 tanks)[8] did not change much during the 1970s. Even given the smaller size of China's infantry divisions, the total number of Chinese forces that may be deployed against Indian forces in the event of war is greater. Regarding the Indian decision to manufacture about 700 Soviet T-72 tanks in India, the pessimists point out that there are similar efforts to acquire even more modern and superior tanks by Pakistan and China. Meanwhile, the type and make of artillery pieces and the semiautomatic Ishapore rifles in the Indian infantry divisions continue to remain the same.[9]

Differing assessments of the traditional sources of threat and the army's ability to meet them are compounded by the new extension of threat perceptions. The more vague and complicated nature of the wider threats now perceived have produced more uncertain and controversial proposals on the army's future role, defense doc-

[8] For an assessment of the performance of Indian armor in 1971 by one of the Indian army commanders, see Singh, *Indian Sword Strikes*, pp. 185–89.

[9] Thapan, *Manpower Ceiling on the Army*, pp. 2–3.

trines, and equipment needs. If potential threats now stem from a wider circle that includes Central Asia, the Persian Gulf, the Indian Ocean, and Southeast Asia, then the army's role would seem to be comparatively subordinate to that of the air force and navy—unless the army is able to include in its functions various airborne and seaborne tasks that are presently being claimed by the other two services.

In particular, the most important of these controversies includes first, the army's claim that it should control all surface-to-air weapon systems (SAMS); and second, the army's proposal for setting up an army air corps that would include fleets of helicopters and transport planes. These two functions are presently controlled and claimed by the air force. In the case of SAMS, the army's case rests on the contention that all weapons used on the ground should be under its control. Traditionally, ground defense against air attack was conducted by the artillery branch of the army with antiaircraft guns. Indeed, at present the army controls Army Observation Post flights that are in support of artillery deployment. Under these circumstances, military commentators sympathetic to the army have suggested that all ground-based air defense should be controlled by one organization, the army. This would produce a more coordinated ground defense strategy and avoid the "wasting of highly trained pilots on static duties."[10] However, the army's claim for control over all aspects of ground-located air defense systems is rejected by the air force for predictable reasons that emphasize the need for vertical integration of all air defense systems from antiaircraft guns and SAMS to fighter interceptors and bombers. This vertical chain of air defense operations is ultimately claimed to depend on the "farseeing eye of the radar."[11] Consequently, the Indian air force has continued to retain command of most SAM and radar systems in India.

The other army/air force dispute is related to the problem of meeting threats in the projected and potential extended theaters of war. The dispute remains at the proposal stage and concerns the formation of an army air corps.[12] A small section of the proposed corps would consist of helicopters, most of which would be for the transportation of infantry and armored divisions. The major part of

[10] Brig. A. C. Cariappa, "Missile Gap in the Indian Armed Forces," reprinted from the *Hindu* in *Strategic Studies* (Islamabad), 3, no. 4 (December 1979):96–98. A response is provided by Group Capt. M. B. Krishnamurthy, "Missiles and Air Defence," also reprinted in the same issue of *Strategic Studies*, pp. 98–101.

[11] Krishnamurthy, "Missiles and Air Defence," p. 99.

[12] See Sinha, *Higher Defence Organisation in India*, p. 17.

the new wing is expected to be composed of airborne divisions and paratroop brigades for possible deployment in extended theaters of war, although this objective is never overtly stated for obvious political reasons. The army has argued for the need to incorporate such heliportable and airborne divisions under its command because these functions usually involve the deployment of ground forces or close aerial support for ground operations for which one central point of coordination would be needed. Especially in its demand for a helicopter wing, the army claims that its pilots could more easily identify the enemy's weapons on the ground and coordinate low-level air operations with ground forces.[13] The helicopter in battle is perceived as a "flying tank" or "antitank" gunship and therefore seems to fall more within the role of the army than the air force. In response to such army proposals, a specially constituted committee recommended in 1978 to the government that a helicopter wing integral to the army be raised. However, because of continued uncertainties and the change of government in 1979–1980, this recommendation has not yet been implemented.

Similarly, controversy over the command and training of seaborne forces for eventual operations on land may arise in the future. Whether or not the Indian army is able to acquire these new functions, or indeed if these functions can ever be realized, it is apparent that in the perceived extended strategic spectrum, the army's role is likely to diminish to the advantage of the other two services.

Strategy and Role of the Air Force

After Independence and partition, six and one-half loosely organized squadrons of the Royal Indian Air Force were assigned to India and the remaining three to Pakistan.[14] This was done according to the communal composition of the RIAF, which was less than one-third Muslim. As noted above, the Indian air force's actions during the first Indo-Pakistani war were for the most part of a noncombat nature, involving reconnaisance and the transportation of troops and supplies to the Kashmir front. The main problem for military operations in the fall of 1947 was that the all-weather road that provided access to Kashmir lay primarily through Murree and Rawalpindi in the new state of Pakistan. The only way to meet the Paki-

[13] See report in the *Patriot*, September 1, 1980.

[14] See Air Marshal M. S. Chaturvedi, *History of the Indian Air Force* (New Delhi: Vikas Publishing House, 1978), p. 58.

stani invading forces quickly was by transporting Indian troops by air to Kashmir. Ground support for the army by the air force was limited because the terrain consisted of high mountainous passes with heavy forestation that concealed the enemy forces and their lines of communication.[15] There were some strafing operations by the air force to relieve the pressure on beseiged Indian army garrisons at Poonch and elsewhere. The Pakistan air force did not enter the war of 1947. Although the reasons for this are not entirely clear, the most likely one is that the PAF was still at the stage of formation, with little organization and experience to embark on a combat role in 1947–1948.

Fifteen years later, Indian air force operations during the disastrous 1962 Sino-Indian war were again almost exclusively confined to supply and transport duties. The major problem faced by the IAF in carrying out these missions was the lack of landing and takeoff facilities in the mountainous regions of the Sino-Indian border. The only two air strips—at Leh in Ladakh and Chusul in the Northeast Frontier Agency—were extremely hazardous; landing and takeoff involved considerable risk. Supply and transport operations therefore had to rely heavily on helicopters or on airdrops.[16] Neither of these tactics was wholly satisfactory because of the vulnerability of helicopters to Chinese groundfire and the inability of aircraft to drop supplies accurately or to guarantee that these supplies would not be damaged on impact.

There are several reasons suggested why the Indian air force did not take on a combat role during this war. First, the unpreparedness and rout of the Indian army, the pride of the Indian services, was felt to be indicative of the likely performance of the air force as well. Air force preparedness for war along the northern borders was even less than that of the army and its entry might have produced an equally dismal fate. As long as the Chinese air force showed no inclination to join the fray, it was considered advisable to confine the IAF to a noncombat role.

Second, the army asked for no supporting military action by the air force. Air Chief Marshal Lal has commented that it was "not that the Air Force was unwilling to join in—indeed the then Chief of Air Staff, Air Marshal Aspy M. Engineer, was most anxious to do so—but apparently the Army did not want it."[17] According to Lal's as-

[15] See Air Commodore A. L. Saigal, ed., *Birth of an Air Force: The Memoirs of Air Vice Marshal Harjinder Singh* (New Delhi: Palit and Palit Publishers, 1977), pp. 236–37.

[16] Chaturvedi, *Indian Air Force*, pp. 122–28.

[17] Lal, *Some Problems of Defence*, p. 68.

sessment, the army until then tended to perceive military operations as primarily its domain with air force support as a kind of "bonus" to be called in if the army considered it necessary. This had been true during the earlier Indo-Pakistani war and was apparently confirmed again in 1962 when the role of the air force was further restricted. "The bare facts of the matter are that in 1962 the Army and the Air Force did not fight together, and whatever operational plans were made were those of the Army alone."[18]

That the Chinese air force did not enter the war indicates operational difficulties. It is now known that Chinese aircraft had difficulty taking off with full fuel and bomb loads from the high-altitude Tibetan bases. Entry of the IAF into the war might therefore have proved advantageous for the beleaguered Fourth Indian Division (of World War II El Alamein fame), which was nearly decimated by the Chinese forces, who outnumbered and outgunned them. At the time it was suspected that the Chinese were reluctant to escalate the war to air combat because they feared retaliation by the United States on the Indian side.

Although the IAF had no combat role in 1962, questions about its role and equipment needs were immediately raised thereafter. During the 1950s, the IAF had been expanded from seven squadrons to fifteen in response to Pakistan's acquisition of aircraft from the United States under the SEATO and CENTO defense pacts. After the 1962 war, the government of India decided to further expand and reequip the IAF into a forty-five squadron force. With this objective, New Delhi approached the United States and Great Britain for high-performance combat aircraft, notably the British Lightning and the American Starfighter, the latter already in service with the PAF. However, the American and British military missions that subsequently visited India concluded that it would be more economical and effective for the western powers to assume direct responsibility for the subcontinent's air defense so that the IAF could concentrate its limited resources on providing tactical support for the army's ground operations against another potential Chinese attack.[19] This offer of a western "air umbrella" was rejected by India.

Problems of coordinating the war strategies of the air force and the army continued during the Indo-Pakistani war of 1965. When the war came, the IAF had reached a strength of thirty-three squadrons out of the forty-five targeted. These were mainly British

[18] Ibid., p. 82.

[19] Saigal, *Memoirs of Air Vice Marshal Harjinder Singh*, p. 287.

Hunter fighters and Canberra bombers, as well as the Gnat interceptor made in India under license from Folland Aircraft and Bristol-Siddley of Britain. However, once again, the army initially entered the war without efforts to seek a joint army–air force battle plan.[20] The role of the air force was expected to be, as in the past, one of logistical support missions with light aircraft. Only after the fighting began was tactical combat support seriously considered, and this occurred when the army found itself under severe pressure in the Chamb-Jaurian sector as described above. At this point Gen. J. N. Chaudhuri was compelled to ask for air cover. Fortunately, Air Chief Marshal Arjan Singh had alerted the air force bases in the Indian Punjab and a force of IAF fighter-bombers from Panthankot mounted a strike immediately.

Even though the air force was for the first time engaged in full-scale combat, its ground-support role was still dependent on the army's strategy. In 1965, the army's need for air support was perceived only belatedly and the IAF therefore operated only on the fringe of the army's military operations. For the most part, the IAF fought a separate air war against the PAF in an attempt to control the skies and to hit each others' strategic targets on the ground.

These discordant purposes of the army and the air force were less evident during the December 1971 Indo-Pakistani war. Much of this was due to a reorganization and synchronization of army and air force command boundaries, put into effect in 1969. Western, eastern, central, and southwestern air force command boundaries were brought approximately in line with the army's western, northern, central, eastern, and southern commands.[21] Much of the air force/army coordination in 1971 also came about because of the prolonged nature of the Bangladesh crisis, which enabled planning by the three services before it erupted into war.

The outbreak of major Indo-Pakistani hostilities in 1971 began when wave after wave of PAF aircraft (consisting mainly of French Mirages) launched a preemptive-style (but unsuccessful) attack on IAF bases at Srinagar, Ambala, Pathankot, Jodhpur, Agra, Utterlai, and Awantipur. The IAF quickly reacted in both the eastern and western theaters. In the east, the IAF knocked out the small PAF stationed there within 48 hours of the start of the war, allowing the army to operate freely without fear of aerial bombardment. In the

[20] See Lal, *Some Problems of Defence*, pp. 69–80; and Chaturvedi, *Indian Air Force*, p. 152.

[21] Lal, *Some Problems of Defence*, p. 82.

west, it provided air cover for the army's operations in Kashmir, Punjab, and Rajasthan. Beyond this, the air force attacked PAF bases such as Sialkot, Sargodha, and Chander, as well as strategic targets around Karachi, Peshawar, and Lahore. An added feature of the air force operations in 1971 was its support for naval operations. In the eastern theater, the liquidation of the PAF by the IAF within two days made it possible for naval vessels and naval aircraft to immobilize the ports of East Pakistan without coming under attack from the air. In the western theater, the air attacks on oil installations and oil fields in and around Karachi were conducted simultaneously with the Indian navy's attack on Karachi harbor. Although Lieutenant General Sinha has written that the naval and air force operations in Karachi were "neither coordinated in terms of their timing nor in terms of their objectives," the concurrent IAF attack, by engaging the PAF, gave the Indian navy more freedom to maneuver off the shores of Karachi.[22]

Just as the army appeared to feel that land operations were its primary domain, Indian air force thinking, derived from its experience since the Second World War, gave primary importance to "winning the air war and establishing air superiority."[23] Support for the army and navy was considered to be secondary to the air force's own battles. The actual experience of 1971 proved to be somewhat different; the air force provided all the support requested by the other two services. Nevertheless, in principle, strategic bombing and air defense operations continued to be perceived by the IAF as its primary objectives.

How does the IAF now see its role in the future defense of India? As in the case of the army, in dealing with extended threat conditions, the IAF faces the problem of "reach." There is also some debate over the degree of urgency or even the existence of threats beyond the traditional defense perimeters. In any case, air force threat perceptions appear to be somewhat more mellow than those of the army and navy. For instance, according to Air Chief Marshal Lal, the traditional Pakistani threat has lessened since the 1971 war both because of the crippling blows dealt during that war and because of the continuous development of the IAF. In an interview, Lal pointed out that Pakistan has only one major port and from it lines of communication for heavy rail and road traffic run in a south-

[22] See Sinha, *Higher Defence Organisations in India*, p. 20.

[23] Lal, *Some Problems of Defence*, p. 80.

north direction, all of which are within easy reach of the IAF.[24] The IAF therefore has the capacity to immobilize the main port of Pakistan and to breach its road and rail lines of communication at several points.

India faces no similar problem. Major Indian lines of communication and most Indian industrial centers, apart from those in the bordering states, are beyond the reach of the PAF. However, the American decision in 1982 to allow Pakistan to purchase the technologically advanced F-16 aircraft would give the PAF an advantage in conducting any future air war. (This advantage was offset between 1982 and 1985 with Indian purchases of French Mirage-2000 and Soviet MiG-29 combat aircraft.) Major industrial centers in Punjab, Haryana, Rajasthan, Gujarat, and Maharashtra (including the heavily industrialized Bombay-Pune sector) would become vulnerable to Pakistani strategic bombing. On the other hand, many other major Indian industrial centers in central, east, and southern India would still be beyond reach. In addition, the IAF, with several forward and rear bases located in the states of Kashmir, Punjab, Haryana, Uttar Pradesh, Rajasthan, and Gujarat, would be in a good position to monitor all PAF activities and thereby to neutralize much of the advantage that Pakistan may acquire through American military assistance.

Lal made these optimistic observations before the Reagan administration took office and based his assessment on military aid that the Carter administration had offered Pakistan; the main combat aircraft was expected to be the F-5Es.[25] The more widespread feeling in the air force was probably expressed by Air Chief Marshal Arjan Singh, although even here the new security environment of India appears to be perceived as less threatening than the other two services consider it to be. According to Arjan Singh, there is little doubt that Pakistan is better off today than before 1971, both because it has a more concentrated area to protect and because of its new military connections with the United States and the Islamic states of the Middle East.[26] If the PAF could perform creditably in the past, it will

[24] Interview with Air Chief Marshal P. C. Lal, January 1981.

[25] In his lecture to the United Services Institution of India in 1977, Lal noted that "I do not think we should be unduly alarmed by what appears to be a strictly commercial deal, for not much can be bought nowadays for the one hundred million dollars that Pakistan is reported to be spending. The more disturbing thing is that Pakistan seems to be forgetting the lessons of 1971 that brought Mr. Bhutto to the conference table" (Lal, *Some Problems of Defence*, pp. 92–93).

[26] Interview with Air Chief Marshal Arjan Singh, December 1980.

do even better with these new military connections. Arjan Singh pointed out that the major constraining factor on the growth of effective striking power in both the IAF and the PAF has been the availability of funds and access to high-performance equipment. Both nations, he believes, have the ability to operate weapons efficiently once they are delivered. Thus, with Arab funding and American and Chinese weapons, Pakistan would have both an economic and a strategic advantage over India. The rate of weapons attrition and cost of replacement make economic considerations equally important in the willingness of either side to go to war.

Similarly, although Lal saw the Chinese threat as in the future more likely to involve "creeping subversion" in Kashmir's Ladakh region and in the volatile Indian states of the northeast rather than outright war, Arjan Singh felt that the experience of 1962 could be repeated in the future at much higher levels of operation that could involve an all-out war between the air forces of the two sides. Other assessments by civilian analysts such as K. Subrahmanyam of the Institute of Defence Studies and Analysis in New Delhi also suggest that the future modernization of the Chinese air force with American assistance and the buildup of new Chinese bases in Tibet including technological improvements in operating aircraft from high altitude bases would make for conditions unlike those in 1962.[27] The IAF may have to contend with a powerful and effective Chinese air force.

Arjan Singh was more cautious and less certain about the role of the Indian air force beyond the traditional perimeters of defense. Although conditions in the Middle East and Southeast Asia may have a bearing on Indian security in a broad sense, responses to these threats by the IAF would have to be in the more immediate vicinity and not in extended theaters of war. However, there have been other Indian assessments that point to the need to consider the growth of the air forces in the Middle East.[28] The procurement of large numbers of Mirage-IIIs/Vs by the United Arab Emirates and Libyan air forces is perceived to have some relevance for the strategic air battles in South Asia. Pakistani pilots train the pilots of these countries and the possibility of the eventual transfer of some of these planes to Pakistan to augment the Mirage fleet of the PAF,

[27] Interview with K. Subrahmanyam, director of the Institute for Defence Studies and Analyses, January 1981. Subrahmanyam was formerly secretary for defense production in the Ministry of Defence.

[28] See Raju G. C. Thomas, "Aircraft for the Indian Air Force: The Context and Implications of the Jaguar Decision," *Orbis* 24, no. 1 (Spring 1980):85–101.

although unlikely, cannot be discounted. Consequently, even if the air war to be fought by the IAF is likely to be in the immediate vicinity, the strategic air balance needs to be assessed in terms of a more extended region.

Whether to limit or extend the role of the Indian air force remains a matter of controversy, but aircraft procurement in the 1980s suggests a trend toward the extended strategic perspective. The American decision in 1982 to sell the advanced F-16 aircraft and other military equipment to Pakistan has already pushed the Indian government into procuring the Mirage-2000 from France and into increasing the number of MiG-25s to be obtained from the Soviet Union. In late 1984 the Indian decision to purchase forty Soviet MiG-29s in order to offset the 40 F-16s of Pakistan immediately and the Pakistani decision to acquire the American E2-C Hawkeye early warning radar-equipped aircraft, which would seriously erode the IAF's role in strategic bombing, suggest a qualitative upgrading of the arms race.[29] This indicates that the role of the IAF will continue to increase in the future.

Strategy and Role of the Navy

Unlike the other two services, the Indian navy had little battle experience under the British in the two world wars and none in the wars of independent India until the Indo-Pakistani war of 1971.[30] The British had thought it best to minimize the development of this service. The southern flank of the Indian empire was protected by the powerful Royal Navy, which, as the cliché goes, had made the Indian Ocean into a British lake. Expansion of the Indian navy even during the Second World War might have raised British fears about access to the Indian Empire itself. Unlike the army, which could be broken up into several regiments, the navy tended to be an integrated unit that could coordinate a rebellion against British rule. This was partly realized in 1946, when sections of the tiny Royal Indian Navy mutinied against their British masters.

Despite the lack of development of an indigenous navy in India

[29] For reports of Indian efforts to acquire the Soviet MiG-29s and Pakistani efforts to acquire the American E2-C Hawkeye, see *NewYork Times*, September 28 and October 12, 1984.

[30] For three related assessments of the Indian navy, see Raju G. C. Thomas, "The Politics of Indian Naval Rearmament, 1963–1972," *Pacific Community* 6, no. 3 (March 1975):452–74; ibid., "The Indian Navy in the Seventies," *Pacific Affairs* 48, no. 4 (Winter 1975–76):500–518; and ibid., "The Indian Navy: Searching for a Role and Rationale," *Marine Policy* 5, no. 4 (October 1981):331–33.

before Independence, the modern Indian navy nevertheless had a long history under the British. Its origin may be traced to 1612 when the British East India Company established a small Indian marine corps at Surat on the western coast.[31] Existing under various titles such as the East India Company's Marine, the Bombay Marine, His Majesty's Indian Marine, and the Royal Indian Marine, this eventually became the Royal Indian Navy from 1934 until 1950, when India became a republic. Under the British, the role of the small RIN was primarily to conduct hydrographic surveys and other auxiliary duties. During the Second World War, the RIN saw some service in the Atlantic and Indian Oceans, mainly on convoy escort duties.

In the partition of British India, the distribution of the RIN between India and Pakistan was approximately the same as in the army and air force except that there was little to distribute, particularly in vessels and equipment. It was only in 1948 that the Indian navy acquired vessels of some caliber: two old destroyers, formerly British, renamed INS *Rajput* and INS *Ranjit*, and a cruiser, INS *Delhi*. After this there were a few similar additions of destroyers and cruisers and the acquisition in the mid-1950s of a 16,000-ton aircraft carrier, formerly HMS *Hercules*, renamed INS *Vikrant*. The strategic purpose of these limited acquisitions was never quite indicated by the government of India, except that the presence of an Indian navy needed to be demonstrated, even if in a token manner.

The navy did not take part in the 1947–1948 Indo-Pakistani war, which was confined to Kashmir, nor in the 1962 Sino-Indian war, which was confined to the Himalayan frontiers, and not even in the expanded 1965 Indo-Pakistan war, although this was conducted on a full scale by the armies and air forces of the two countries. However, during the 1965 war, the Indian navy was placed on alert and there was at least one minor naval engagement at Dwarka on the Kutch coast, which was bombarded by the Pakistan navy. Despite the Indian navy's nonparticipation, some reports indicate that there might have been an Indonesian naval threat during the war. President Sukarno, who had recently acquired a $1 billion navy from the Soviet Union, had made vague threats to intercede on the Pakistan side. Later, Pakistan's Air Marshal Asghar Khan revealed in his memoirs that Sukarno had in fact offered to divert Indian attention

[31] See the essays by the former chief of naval staff, Adm. S. N. Kohli, in his *Sea Power and the Indian Ocean* (New Delhi: Tata McGraw-Hill, 1978), pp. 13–20. See also Vice Adm. M. P. Awati, "Indian Navy-I: An Ocean of Problems," *Patriot*, December 7, 1979.

from Pakistan by seizing the Andaman and Nicobar islands in the Bay of Bengal.[32]

It is not clear why the Indian navy was not involved in 1965, but this may be attributed to the lack of a Pakistani naval threat, the inexperience and the small size of the Indian navy at the time, and the fact that the army had not requested supporting naval operations. Whatever the reasons, naval demands for a greater share in the development of the services surfaced from 1966 on under two successive chiefs of naval staff, Admirals A. K. Chatterji and S. M. Nanda. Three basic arguments for naval expansion were advanced. First, the naval lobby argued that the development of the navy should depend primarily on the geostrategic location of the country and the nation's perception of its role as a naval power in the region.[33] Naval growth did not have to be directly related to immediate and clearly specified threats in the vicinity. The rationale underlying this argument was the long time span needed to develop and nurture a navy. Unlike infantry or even armored divisions, naval capabilities could not be raised within a short period. As such, the ability to meet sudden and specific naval threats could only be undertaken on the basis of long-term planning.

Second, the advocates of naval expansion pointed out certain economic and protective purposes in the role of the navy. According to this argument, the navy was also concerned with the protection of India's seaborne trade and offshore resources. Continuing growth in this area of the Indian economy necessitated the continuous and parallel growth of Indian naval power. Third, proponents of the navy's growth spoke of the need for some degree of military balance among the three services so that strategic assessments were not biased by a dominant service. In opposition to these arguments, the strongest case against naval expansion in this period was the lack of a visible naval threat from Pakistan, China, or any other regional power.

Arguments for limited expansion carried some weight among Indian defense planners. From the mid-1960s on, there was modest but steady acquisition of combat vessels by the Indian navy, primarily from the Soviet Union. These proved fortuitous in 1971, when the Indian navy participated in a major war for the first time. Although the performance of the Indian navy appeared spectacular—

[32] See Kohli, *Sea Power and the Indian Ocean*, pp. 133–34.

[33] See, for example, Kohli, *Sea Power and the Indian Ocean*, pp. 21–34; and Adm. A. K. Chatterji in "The Navy," in Press Institute of India, eds., *Defence of India* (New Delhi: Vikas Publications, 1969), pp. 21–24.

the small Pakistani navy was easily crushed—the results had as much to do with the uneven sizes of the two navies as with the well-planned Indian naval tactics under the chief of naval staff, Admiral Nanda.

In the eastern theater, there was virtually no Pakistani naval opposition. The INS *Vikrant*, with its Seaking fighter planes and accompanying destroyers and frigates, was easily able to sink almost all the merchant ships in the main ports of east Pakistan at Chittagong, Khulna, and elsewhere that could have been used for escape by the Pakistan army. A major Pakistani loss of a combat vessel occurred in the east when the submarine *Ghazi* was sunk early in the war near the port of Visakhapatnam. In the western theater, the Indian navy surprised the Pakistani naval and air force units defending the port of Karachi. Due to their short travelling range, the Soviet-supplied Osa missile boats stationed at Bombay were not expected to pose a threat; but these boats, to the surprise of Pakistan, were towed more than halfway for the raid on Karachi.[34] The naval raid, which took place at night and had withdrawn beyond the range of the Pakistan air force by dawn, was greatly assisted by the IAF, which simultaneously engaged the PAF in an air war over Karachi.

The success of Indian naval operations in 1971 again raised the question of naval threats to India; the defeat of the Pakistan navy meant the elimination of the only naval threat visible. However, developments in the 1970s have significantly strengthened the case for naval expansion, especially with respect to the other two services. Much of the argument conforms to the points advanced in the late 1960s, which by the 1970s had begun to appear self-evident. These proposed functions may now be grouped under three basic categories: a "blue water" offensive and defensive military role against perceived opposing naval powers in the extended strategic environment; a coastal defense role within some designated defense perimeter around the Indian peninsula; and a defensive role intended to protect Indian shipping in the coastal waters and on the high seas.

In the first case, expansion of naval forces is largely justified by

[34] In an interview in December 1980, Adm. S. M. Nanda, explained that Karachi was well-guarded by Pakistani vessels strung around the port about fifteen miles apart. The Indian naval flotilla lay bunched together off the Kutch coast, drawing the curiosity of the Pakistani vessels, which set out to investigate the situation one at a time. Four or five Pakistani ships were destroyed until the Indian naval ships outnumbered the Pakistani ships. A full-scale attack was then launched at night by the Indian navy. Interview with Admiral S. M. Nanda, December 1980.

the increasing relevance of various arms races taking place in the Middle East, the Indian Ocean, and Southeast Asia. Although most of the arms accumulations, especially in the Middle East, are land- and air-based systems, Iran under the Shah, had sought to build a powerful naval force in addition. As already noted, Indonesia under Sukarno had also attempted to develop a powerful naval force. More recently, the Shah and his ambitions of becoming a major regional power gone and Iran in disarray, the weapons acquisitions of Saudi Arabia, Oman, Iraq, Egypt, and Libya have taken on greater significance. These weapons do not appear to be directed against India but to stem from military rivalry in the region; danger to India arises from the possible transfer of weapons to Pakistan. Under these circumstances, an extended naval combat role would occur only if India extended the arena of conflict to those states in the Middle East or Southeast Asia that might be actively assisting its adversaries on the subcontinent.

Of greater naval relevance is the superpower rivalry in the Indian Ocean. The Indian navy clearly cannot hope to compete with these navies. However, Admirals Chatterji and Nanda have suggested that a powerful Indian navy could at least harrass superpower navies and raise the costs of their intervention in the region of the Indian peninsula.[35] Without Indian naval opposition, interventionist threats by the superpowers may occur more often and appear more credible. Some resistance by India would at least raise the stakes for superpower intervention. Such a situation is claimed to have existed during the 1971 Indo-Pakistani war when the USS *Enterprise*, with accompanying escort ships, entered the Bay of Bengal.

Such perceptions may appear fanciful and may be seen as arguments to rationalize naval demands for the development of a "blue water" navy. Nevertheless, such scenarios are also complemented by more pragmatic perceptions that largely revolve around the Pakistani naval threat. The immediate question is whether attacks on the Pakistan navy should be confined to Karachi harbor as during the 1971 war, or whether future hostilities should call for the interdiction of Pakistani shipping on the high seas to weaken Pakistan's ability to continue the war.

The second major role of the Indian navy stems from the increasing importance of coastal defense. Much of this is due to the extension of territorial waters to two hundred miles and the search for offshore mineral resources, especially oil. The discovery of substan-

[35] Interviews with Admirals S. M. Nanda and A. K. Chatterji, December 1980.

tial quantities of oil and natural gas in the Bombay High, off the western coast, has led to further exploration in the southern and eastern offshore areas of India. Increases in fishing and coastal commercial traffic have added to the need for coastal defense. These developments prompted the government to set up a coast guard in 1978, a service separate from the navy.[36] The new organization is being equipped with seaward defense boats and fast patrol boats. Although it constitutes a separate service, it is expected to function under the command of the navy during times of crisis.

The third role derives from the growth of India's shipping and seaborne commerce, which increased from 200,000 gross registered tonnage (GRT) in 1947 to 5.5 million GRT in 1978 and in value from RS. 1,800 crores ($2.2 billion) in 1960 to RS. 13,000 crores ($16.25 billion) in 1979.[37] Although the actual protection of Indian shipping over long distances would be costly and difficult if not impossible, this third function emphasizes the need to protect Indian shipping lanes in potential battle zones where India may be involved in the future.

Notwithstanding the scenarios based on the extended strategic framework, at present the Indian navy does not have an effective "blue water" capability for the conduct of war in theaters beyond the subcontinental waters, the interdiction of Pakistani shipping, or the protection of Indian shipping on the high seas. It has essentially a "brown water" capability.[38] With the exception of the aging carrier *Vikrant* and some Kashin-class destroyers supplied by the Soviets, which may operate to a limited extent beyond coastal waters, the navy's Soviet Foxtrot-class and German Type-209 submarines, Indian-made Godavari-class and Soviet-built Petya-class frigates, and Soviet-supplied Osa-class torpedo boats and Nanuchka-class corvettes are intended for coastal defense.

Overall, the extended strategic framework essentially points to a greater increase in the role of the Indian navy in the future in comparison with the other two services, although the Navy's declaratory policy has not yet been matched by the government's action policy. Unlike the army and air force, where the ability to conduct wars in more extended areas is more problematic because of restrictions on mobility or the range of their striking power, the growth of the navy

[36] See annual *Report: 1979–80* (New Delhi: Ministry of Defence, Government of India, 1980), pp 21–23.

[37] See Raju G. C. Thomas, "India's Defence Posture in the 1980s," *Hindustan Times*, January 10, 1981.

[38] See Bharat Karnad, "An Optimized Navy," *Hindustan Times*, June 9, 1984.

tends to improve such mobility and range. Moreover, even if strategic developments in more extended areas still appear somewhat irrelevant for Indian defense today, the navy can justify expansion based on the need to defend the extended Indian territorial waters and the growth of seaborne commerce.

The Coordination of Defense Roles under the Extended Strategic Framework

The often disparate and uncoordinated operations of the army, air force, and navy during the four major wars since Independence and the present broader perspective of India's strategic environment have prompted proposals for a more unified strategy by the three services. In the 1971 war a fair amount of coordination had been achieved, but this, as noted earlier, was due to the presence of compatible military and civilian leaders rather than organizational structure and planning. A more coordinated defense system has also been suggested since then because of the steady expansion of the services. India has become a major regional power, which calls for an appropriate reorganization of its defense system and revisions in strategic planning. Additionally, with the increasing prospects that India might opt to become a full-blown nuclear power, establishing new strategic doctrines to balance India's conventional and nuclear capabilities is imperative.

Although all three services do not deny the need for such strategic rethinking and military reorganization, the two smaller services continue to be apprehensive about the effect of change on their inputs into the policy-making process and their relative growth with respect to the present dominant service, the army. As discussed in chapter four, proposals for creating a chief of defense staff, for instance, have usually been supported enthusiastically by the army because the system would enable inter-service coordination and joint strategic planning. However, opponents of the CDS system in the air force and navy, such as Air Chief Marshals Arjan Singh and Lal and Admiral Nanda, argue that from a strategic standpoint a CDS becomes necessary only when dealing with overseas theaters of war in which all three services are engaged. As long as wars are expected to be fought along the borders of India, sufficient coordination may be achieved through the present inter-service arrangements that are directed from a central point in New Delhi.

Because the extended strategic framework suggests potential overseas inter-service operations, the CDS issue may be revived later. Meanwhile, Lt. Gen. M. L. Thapan, in a report to a seminar at the

United Services Institution of India, proposed an integrated set of command systems at lower levels of the army, air force, and navy defense organizations. Working directly under a CDS or a similar system, the defense of India, according to Thapan, could be entrusted to five theater commanders:

(a) Northern Command, with headquarters at Chandigarh, with an army commander in chief and deputies from the army and air force;

(b) Central Command, with headquarters at Lucknow, with an air force commander in chief and deputies from the army and air force;

(c) Western Command, with headquarters at Ahmedabad, with an army commander in chief and deputies from all three services;

(d) Eastern Command, with headquarters at Calcutta, with an army commander in chief and deputies from all three services;

(e) Southern Command, with headquarters at Bangladore, with a naval commander in chief and deputies from all three services.[39]

The proposal leaves three regional command posts under army chiefs and one command post each under air force and naval chiefs. Army and air force deputies would exist in all five commands but naval deputies would exist in only three. This arrangement presumably could be modified through mutual consultation among the three services. According to General Thapan, the scheme would produce coordinated military planning as well as common economic benefits. For example, at present the army's Western Command is located at Simla, that of the air force at Delhi, and the navy at Bombay. Even if the integrated theater concept required more time to develop, Thapan felt that the common location of command headquarters of the three services was needed immediately to ensure meaningful inter-service cooperation. He stated further that "the present method of coordinating the activities of the three Services was a hit or miss method and particularly unreliable in peacetime when defense preparations should be at their peak."[40] In particular, the fortunate circumstances of 1971 could not be expected

[39] Lt. Gen. M. L. Thapan, *Cooperation in Defence* (New Delhi: United Services Institution of India, 1978), pp. 8–9.

[40] Ibid., p. 8.

to prevail in the future as defense organization and strategy became more extensive and complicated and the size of the services increased dramatically. Thapan concluded that "in today's context, defence [is] much more than the sum of the three Services."

Such proposals for a coordinated inter-service defense organization have yet to be implemented. The reasons, noted above, have been due in part to civilian fears of coordinated military takeovers and in part to fears within the two smaller services that they might be dominated by the army. However, as civil-military cooperation and confidence grows in India and as the Indian air force and navy increase in size and confidence, the advantages of coordinated military systems are likely to be recognized and thereafter translated into reality.

The Role of the Services in Internal Security and National Politics

An important issue in many Third World countries is the role of the armed forces in national politics and, more specifically, in the maintenance of internal security.[41] Armed forces are normally raised to defend against external threats. But their size and national visibility often make them a considerable force—whether actual or latent—in domestic political affairs. The use of regular military forces to deal with internal as well as external security has tended to increase the influence and power of the military in domestic affairs; often this has led to military takeovers. Similarly, in India, the new concept of security that broadly encompasses problems of internal security and domestic economic and political stability in addition to defense preparedness against external threats has caused concern about the political influence of the armed forces and the effect of this on the democratic system.

Generally, forces and programs that deal with internal unrest or uprisings are kept—or ought to be kept—separate from forces intended to oppose external threats. This has usually been the case in western democracies and has been true also of the Indian democ-

[41] Some prominent books that deal with this issue are Samuel E. Finer, *The Man on Horseback: The Role of the Military in Politics* (Baltimore: Penguin Books, 1976); Samuel P. Huntington, *The Soldier and the State: The Theory and Politics of Civil-Military Relations* (Cambridge: Harvard University Press, 1957); Amos Perlmutter, *The Military and Politics in Modern Times* (New Haven: Yale University Press, 1977); Morris Janowitz, *The Military in the Political Development of New Nations* (Chicago: University of Chicago Press, 1964); and Gavin Kennedy, *The Military in the Third World* (London: Duckworth Press, 1974).

racy. However, in most parts of the developing world, the presence of authoritarian regimes—particularly military dictatorships—has tended to blur the distinction between internal and external security. Security problems at the domestic level may be linked to problems at the international level and the same forces and equipment may be used for internal and external security. An example is the suppression of the separatist movements in Bangladesh and Baluchistan by the Pakistan army. These forces are used also—as they are intended to be—to meet threats from Afghanistan and India. Similarly, the Indian army has been used to suppress the separatist movements in Nagaland and Mizoram as well as to meet the Chinese threat across the Himalayan frontiers. Some military or military-backed civilian regimes use their armed forces primarily for the maintenance of internal security. This may make sense where the potential for internal revolutions and armed insurrections is perceived to be greater than the chance of external attacks on the state. Other regimes normally maintain internal security through regular and separate police forces. The army enters the picture only when internal problems become severe and the police, more lightly armed and fewer in number, are unable to handle events—a situation that is not fundamentally different from that of India.

In India, however, civilian authorities control the military and limit its role in the security policy-making process. Although there are military inputs into the formulation and conduct of external defense policy, basic doctrines and policies are developed by the elected civilian government in power and ultimate authority rests with it. In the case of internal security, the armed forces do not have any formal participation in the security policy-making bodies, although they have been used in the maintenance of internal security.

Although the growth of paramilitary forces in India (300,000 in 1983 as opposed to 1.3 million in the armed forces) may cause concern among some civilian leaders, the armed forces in India prefer this arrangement, provided the lines of authority are clearly drawn between external military and internal paramilitary roles. For example, border security forces and coast guards must ultimately operate under the supervision of the armed services even if some of these forces fall under the jurisdiction of the Ministry of Home Affairs rather than the Ministry of Defence.

The Indian army has also accepted the inevitability of their operations against armed guerrillas in states like Nagaland, Mizoram, Meghalaya, and Tripura in the northeast sector. However, their deployment in the country against terrorist activity or to control

Hindu-Muslim or other forms of civilian violence is generally not welcomed by leaders of the armed services. Despite the precedent that existed under British rule, when the Indian army under British officers was used for internal security purposes, the leaders of the post-Independence armed forces feel that in a free and democratic country, the military should not be used against its own people. The armed services constitute one of the major and most visible symbols of national unity in a country of extreme diversity where divisive movements are perennial. Even if effective, it is only with reluctance that the Indian army has played a role in curbing Sikh terrorism in the Punjab or communal violence throughout the country.

Certain civilian roles have been advocated for the armed services during times of peace. Given the considerable size of the Indian armed services and the paramilitary forces, suggestions have been made that this manpower be utilized in development projects during prolonged periods of peace. An alternative suggestion has been that a large "people's liberation army" similar to the Chinese model be added to the regular military. This would channel the millions of unemployed or underemployed in the country into more productive uses while perhaps enabling a reduction in the size of the regular armed forces. Either way, the large and inflationary cost of maintaining a sizable military in India—the fourth largest after the Soviet Union, China, and the United States—would be mitigated.

There have been objections to such proposals from both politicians and the military. Politicians fear that a massive organization like a people's liberation army could potentially carry out a military takeover of the country. The armed services question the utility of such a large force in the high-technology wars of the present. At the same time, they are against the use of the regular military in development projects during times of peace, because they believe this would erode the professionalism and combat effectiveness of the armed services during times of war. The military accept, however, their civilian role in disaster relief and in the manning of basic services during prolonged nationwide strikes in key industries and services.

II. Defense Planning and Programs

The transformation of the Indian strategic environment in the 1970s produced various proposals for defense programs, and these began to be taken seriously by the government toward the end of the decade. Initially, adjustment to the new environment was made through modernization programs. Much of the equipment of the services, especially in the air force and navy, was based on the technology of the 1950s and was due for replacement. However, the quality of some of the new equipment proposed for acquisition suggested not merely replacement but a major increase in Indian defense capabilities. Again, though little has been said about the underlying strategic purposes of these weapons other than describing them as replacements for aging equipment, since 1975 the government has continuously expressed concern about strategic developments beyond the subcontinent that may have relevance for Indian security.

Until recently, reasons for the gap between the extended concerns India expressed and the more limited defense programs it proposed were due in part to the ambiguities inherent in the new threats perceived and in part to the inevitable time lag that occurs in moving from threat perceptions to appropriate defense programs. Programs already under way are difficult to reverse, while newer programs that upset the prevailing bureaucratic structure are likely to meet with resistance. Moreover, the possibility exists that newer threats may be transitional; a hasty defense response may consolidate a threat that might otherwise have merely passed. These factors are further complicated by the more heated debate on whether India should become a nuclear power. To opt for nuclear weapons would, of course, affect India's traditional defense postures and call for an examination of what constitutes an optimal distribution between conventional and nuclear weapons. Would the acquisition of nuclear weapons, for instance, call for a lower level of conventional weapons or would it leave the conventional program unaffected?

Force Levels and Defense Programs of the Army

Army defense programs continue to be relatively simple and easier to execute than those of the other two services. Since the last war in late 1971, there has been an increase in men under army uniform from 830,000 to approximately 1.1 million in 1985—an increase of 32 percent. This compares with an increase of 23 percent in Pakistan, from 365,000 to 450,000.[42] The growth of both armies was also manifested in an increase in their number of infantry divisions. In India the army went from thirteen infantry divisions to nineteen by 1985; in Pakistan, from twelve infantry divisions to sixteen. On

[42] Figures from *The Military Balance, 1971–72* and *1985–86* (London: International Institute for Strategic Studies, 1971 and 1985).

either side, the number of armored divisions remained at two apiece, with independent armored brigades increasing from three to seven in India and from one to four in Pakistan in the first decade after the 1971 war. In 1985, about three of the sixteen infantry divisions in Pakistan were deployed along the Afghan border, and about four of the nineteen Indian infantry divisions were deployed in the northeast, where India must take into account five infantry divisions in Bangladesh and counter various armed insurgencies in the tribal states bordering Burma and Bangladesh. This leaves fifteen infantry divisions, two armored divisions and probably three to four armored brigades that India and Pakistan each have directed primarily against each other.

Against China, the ten mountain divisions that were raised in the immediate years following the 1962 war remained the same in 1985. These divisions were directed against twenty-six infantry divisions deployed by China along their borders in South Xinjiang, Xizang (Tibet), and Yunan. Since South Xinjiang and Yunan provinces primarily face Afghanistan and Vietnam and because a Chinese infantry division consists of about 12,700 men compared to 17,500 in an Indian division, the direct confrontation of Indian and Chinese forces in the early 1980s was approximately equal.

Thus, except for Bangladesh, from which no serious military threat is anticipated, India has maintained parity on the ground with Pakistan and China, division for division and man for man. However, India's mountain divisions are infantry divisions acclimatized to live at high altitudes. They therefore can be diverted to the Pakistani front as was done during the 1971 Indo-Pakistani war; three to four mountain divisions from Arunachal Pradesh (formerly the North East Frontier Agency) joined three infantry divisions for the invasion of East Pakistan. The consolidation of Pakistan in one contiguous territory today does not negate the possibility that this could occur again. The mountain division deployed along the Ladakh-Tibet border could quickly be moved to the Kashmir or Punjab fronts. Nevertheless, such a strategy would upset the balance of ground forces against China and open a gaping hole in India's defenses along the Himalayan borders. Such a calculated risk was probably taken in December 1971 on the basis of the possible deterrent value of the freshly signed Indo-Soviet Treaty of Peace and Friendship of August 1971. New Delhi may have also assumed at the time that China's stake in the Bangladesh issue was not critical enough to warrant PRC mililtary action against India. These circumstances may not prevail in another subcontinental war.

The other prospect of an imbalance in ground force levels between India and Pakistan would occur if—as according to some unconfirmed reports—Pakistan were to station two infantry divisions in Saudi Arabia while two compensating divisions were raised in Pakistan, equipped and maintained at Saudi expense.[43] Like India's mountain divisions facing China, however, this would not immediately upset the balance of ground forces along the Indo-Pakistani border; if a sudden confrontation took place, it would be several weeks before Pakistani divisions stationed abroad could be returned to the Indian front. (As of 1985, there was no evidence that Pakistani divisions have been stationed in Saudi Arabia.)

Given the balance of troop distributions between India and its neighbors, especially with reference to Pakistan and China, imbalances are likely to occur mainly in the quantity and quality of weapons carried by India, China, and Pakistan. Against both the traditional adversaries, Indian army commanders have always stressed the need to possess a qualitative balance of equipment, especially in light of the short wars that have been fought. In this respect, it is relevant to note in the Indian case that the government of India has thus far imposed a manpower ceiling of one million for the Indian army.[44] This limit obviously has little to do with a lack of manpower, of which there is an abundance in India; it had more to do with strategic assessments made immediately after the Sino-Indian war when a total complement of one million men was targeted as a desirable round figure to be attained by the army.

In the mid-1980s, the ceiling has been exceeded only marginally in part for economic reasons; the bulk of the expense of the army consists of wage payments and the costs of provisions and stores. Another reason was to induce more efficient use of trained manpower. Strategic assessments perhaps also did not suggest a need for any upward revision of army manpower. In addition, a round figure like one million carries a psychological rationale in itself and is not easily dislodged. Whatever the reasons for the present limit, any major increase in the army's defense capabilities must therefore come through improvements in the quality of its front-line equipment.

The traditional needs of the army are tanks and anti-tank weapons, including short-range missiles for use on the battlefield; armored personnel carriers (APCs); artillery; rifles and ammunition;

[43] *New York Times*, February 6, 1981.

[44] See Thapan, *Manpower Ceiling on the Army*.

and transport vehicles, including planes and helicopters. In the context of the new roles perceived to lie ahead, the army's need for airborne transport has taken on an added dimension and has been expressed in the demand for an army air corps.

Tanks have constituted the central equipment of India's two armored divisions and five independent brigades. At the time of the 1971 war, the main tanks in service with the army were 459 T-54/55 and PT-76 tanks purchased from the Soviet Union in the 1960s; 200 Centurion and 100 AMX-13 tanks purchased from Britain and France in the 1950s; 300 Vijayanta tanks (of the British Chieftan design) made in India under license from Vickers-Armstrong, and about 250 American Sherman tanks obtained during the Second World War. By 1981, almost all of the Sherman, the Centurion, and the AMX-13 and PT-76 tanks had been withdrawn; only a few of the last two items still remained with some of the armored brigades. In 1985, the principal tanks consisted of 700 Soviet T-54/55 tanks and 1,500 Indian-made Vijayanta tanks. This was viewed as unsatisfactory, mainly because the technology of the existing tanks was considered to be obsolete. One report indicated that the Vijayanta tank's performance during the 1971 war did not come up to expectations. According to Brig. A. C. Cariappa, the army subsequently "was not confident enough to equip its armored divisions with an indigenous tank."[45] Instead these divisions were equipped for the most part with the Soviet T-54/55 tanks; the Vijayanta "was relegated to the infantry divisions as anti-tank protection weapons which any ordinary gun or missile should accomplish with much less cost."[46]

Since both the remaining Soviet and Indian tanks were primarily based on the technology of the 1950s, selecting a new battle tank for the army had preoccupied the government since the mid-1970s. Under the Janata government, three tanks were considered for purchase and eventual manufacture in India. These were the British Chieftan MK-2, the German Leopard-2 and the Soviet T-72.[47] Initially, some advantages were seen in the British Chieftan-2 in that the Vijayanta had been based on the earlier version of the tank and several hundred were obtainable cheaply because they had already been manufactured for the now-deposed Shah of Iran. On

[45] Brig. A. C. Cariappa, "The Choice of Main Battle Tank," *Hindu*, April 23, 1980; See also Brig. R. D. Law, *A New Battle Tank for India* (New Delhi: United Services Institution of India, 1975).

[46] Cariappa, "The Choice of Main Battle Tank," *Hindu*, April 23, 1980.

[47] *Hindustan Times*, May 11, 1980.

the other hand, the Leopard-2 was considered to be the most superior tank available in the West European arms market. Neither, however, compared favorably in price or production terms with the Soviet T-72. Additionally, the control equipment and some other parts of the Leopard-2 were manufactured in the United States, which made political intervention by Washington a possibility. This belief was based on earlier Indian experience with efforts to acquire the Swedish Viggen aircraft, which was powered by an American Pratt and Whitney engine. Ultimately, Indira Gandhi's Congress government chose in 1980 to acquire the Soviet T-72 tanks, to be purchased initially and later manufactured under license in India. By 1985, 300 T-72 tanks had been purchased and another 1,600 were on order.

Until 1980, the main antitank guided weapons (ATGWS) of the army were the French SS-11s and the ENTACS manufactured under license at Bangalore.[48] The ATGWS were supplemented by 57mm and 107mm antitank guns, items that had been in service for more than a decade and were overdue for replacement. In a major attempt at a shift in policy in 1980, India sought to acquire from the United States about $40 million worth of TOW antitank missiles, consisting of 60 launchers and 3,500 missiles, and also 230 light-weight, long-range M-198 howitzers worth about $300 million to equip its mountain divisions.[49] But negotiations with the United States proved inconclusive for both economic and political reasons. According to the Indian version, the terms offered by the United States to India called for advance cash payment in full. However, delivery would be made over a period of several years, with the United States holding the right to veto delivery at any time if warranted by subsequent political circumstances. Although these negotiations fell through, Pakistan concluded an agreement with the United States in 1981 for the procurement of an unspecified quantity of TOW missiles. Subsequently, the Pakistan government was also able to obtain several hundred M-48A tanks and armored personnel carriers from the Reagan administration.

Two other major items of equipment for the army are guns for its fourteen independent artillery brigades; and APCS, rifles, and ammunition for its infantry divisions. All the 25-pounder guns of World War Two vintage in the artillery regiments were phased out

[48] From *The Military Balance, 1980–81*.

[49] See the *Hindustan Times*, October 18 and 19, 1980; and the *Statesman*, October 14 and 19, 1980.

in the 1970s and replaced by a standard Indian-made 105mm gun.[50] Similarly, the air defense capabilities of the artillery were enhanced by replacing the older antiaircraft guns with indigenous L-70 guns. As regards APCS, there were 500 BTR-60/142s and OT-62A/64As in armored and infantry divisions in 1985. More APCS in the army, both tracked and wheeled vehicles, were expected to enhance the mobility of the infantry divisions in battle.[51] APCS were considered simple enough to be manufactured quickly: the wheeled type at the Shaktiman truck factories and the tracked type at the Heavy Vehicles Factory at Avadi where the Vijayanta tanks are manufactured.

Somewhat more controversial than proposals for an increase in APCS to mechanize the infantry divisions were calls for setting up a helicopter wing. Disagreement arose over the vulnerability of heliportable divisions, especially near battle sectors, and from the unresolved question of whether helicopters should be flown by air force or army pilots. Meanwhile, the army operated some forty Cheetah (formerly Alouette-3) helicopters in 1983 made at the Hindustan Aeronautics Ltd. (HAL) in Bangalore under license from France. Another seventy-five were on order from HAL and were in the process of being delivered in 1985.

Force Levels and Defense Programs of the Air Force

The air force is the second largest wing of the armed services in India. Even then it remains a junior partner to the army. The average annual resource allocations to the air force have been around 20 percent of the total defense budget compared to the army's 65 percent and the navy's nearly 10 percent (the other 5 percent is allotted to service pensions). Nevertheless, air force defense programs tend to be more visible and controversial than those of the army. Air force programs carry a greater dependence on foreign manufacturers and consequently also on foreign exchange. Aircraft technology is relatively more complex, the destruction of aircraft in battle or even in accidents involves sizable financial losses that are not easily recovered, and combat aircraft in general tend to have higher rates of obsolescence than the equipment of the other two services.

At the time of the 1971 Indo-Pakistani war, the Indian air force had reached the full complement of forty-five squadrons that had

50 *Report. 1979–80*, Ministry of Defence, p. 14.

51 For a set of Army seminar reports on this issue, see Maj. Gen. D. Som Dutt, *Armored Personnel Carriers* (New Delhi: United Services Institution of India, 1973).

been recommended by the Tata Aeronautics Committee immediately after the 1962 Sino-Indian war.[52] This included seven interceptor squadrons of MiG-21s and eight of Gnats/Ajeets; five fighter-bomber squadrons of Sukhoi-7Bs, two of HF-24 Maruts, six of F-56 Hunters, and two of Mystère-4s; three light bomber squadrons with B-1 Canberras; and several other reconnaissance and transport squadrons consisting of an assortment of Canberra PR-57s, C-47s, C-119s, Ilyushin-14s, Anotov-12s, HS-748s, Caribous, and Otters.[53] There were also some helicopter squadrons of Mi-4s, Alouette-3s/Cheetahs, Bell-47s, and a few Mi-8s. The air force also controlled fifty Soviet supplied SA-2 SAM complexes.

On the combat side, the Canberras, Hunters, Mystères, and Sukhois were purchased directly from Britain, France, and the Soviet Union. The MiG-21s and Gnats were made under license in India in collaboration with the Soviet Union and Britain. By the late 1960s, the Gnats had become wholly indigenous after the expiration of the licensing period and were renamed Ajeets. The airframe of the Marut was designed in India with German assistance and its engine is manufactured under license from a British company. On the transport side, almost all the aircraft were directly purchased from abroad except for the HS-748 transport planes and the Alouette-3 helicopters, which are made in India in collaboration with Britain and France. All of these aircraft are now almost wholly indigenous.

Except for the Sukhoi-7Bs, which were negotiated and purchased from the Soviet Union in the 1960s, all other aircraft were either purchased, manufactured, or negotiated for manufacture before the 1962 Sino-Indian war. Subsequently, there were some incremental technological upgrading of the aircraft at various times, especially their electronics and missile capabilities as in the case of the MiG-21s. However, the basic technologies of all these planes belonged to the 1950s. Even the Sukhoi-7B was based on Soviet technology developed in the 1950s. By the 1970s, almost the entire Indian air force was due for modernization.

To a certain extent, this situation paralleled that of the Pakistan air force, whose F-104 Starfighters, F-86 Sabres, and B-57 Canberras supplied by the United States were also ready for replacement. Similarly, the Chinese-supplied F-6/MiG-19 fighters of the PAF, like the SU-7Bs in India, were based on obsolete technology. The exception in the Pakistan case was the acquisition of the Mirage-3/5

[52] See Raju G. C. Thomas, *The Defence of India: A Budgetary Perspective of Strategy and Politics* (New Delhi: Macmillan Co. of India, 1978; Columbia, Mo.: South Asia Books, 1978), p. 191.

[53] From *The Military Balance, 1971–72*.

fighter-bombers from France in the late 1960s through the 1970s.[54] Despite India's continued numerical superiority in combat aircraft over the last three decades, Pakistan was considered to have had a substantial qualitative superiority in 1965, given its recently acquired American aircraft, and marginal qualitative superiority during the 1971 war because of its Mirage acquisitions. In the mid-seventies, fears were expressed that the Pakistan Mirage fleet of some ninety aircraft might be augmented by the transfer of similar aircraft from the Libyan and the UAE air force, where the Pakistan air force has provided training for the pilots. These conditions, together with the obsolescence of the earlier aircraft purchased in the 1950s and the perceived technological superiority of the Pakistan Mirages, prompted serious Indian discussion for acquiring and later manufacturing under license a new "deep penetration strike aircraft."

Initially, four basic aircraft options were considered: the Soviet MiG-23, the Swedish Saab-37 Viggen, the French F-1 Mirage, and the Anglo-French Jaguar. In September 1979, the Jaguar was selected over the others by the Janata government. The MiG-23 had been eliminated earlier on the grounds that it did not meet the IAF's specifications for a deep penetration strike aircraft, although more likely the decision may have stemmed from the Janata government's desire to reduce India's military dependence on the Soviet Union. The Swedish Viggen was eliminated mainly because of a threat by the United States to veto the sale since the craft was powered by an American Pratt and Whitney engine. According to the Secretary of State, Cyrus Vance, the Swedish sale would indirectly amount to a violation of an American arms embargo to the subcontinent and upset the military balance between India and Pakistan.[55] The final choice of the Jaguar over the French Mirage was largely based on cost and delivery schedules. The total Jaguar price was projected to be $300 million less than that of the Mirage, while the British also offered to divert immediately some Jaguars from the Royal Air Force to the IAF after they had been suitably tropicalized. To counter the French offer to share the technology of the Mirage-2000 in the future, the British offered to do the same with the Tornado, the planned successor of the Jaguar.

In an effort to replace the Canberras and Hunters and partially meet the Pakistani acquisition of Mirages, the Indian decision to ac-

[54] See Thomas, "Aircraft for the Indian Air Force."

[55] Ibid., p. 95. The threat by Secretary of State Cyrus Vance to veto the Swedish sale of Viggens to India was reported by Inder Malhotra in the *Times of India*, September 2, 1978.

quire the Jaguar may in fact have resulted in the IAF achieving clear superiority over the PAF. From a state of technology that seems somewhat behind the technology of Pakistan's best aircraft, the Jaguar has in effect "leap-frogged" Pakistan's Mirages at considerable expense to the Indian treasury. This led to frantic Pakistani efforts to offset the new Indian aircraft acquisition, which met with success eighteen months after the Soviet invasion of Afghanistan when the Reagan administration decided to sell forty of the advanced F-16 fighters in its package arms deal of $2.5 billion to Pakistan.[56] Unfortunately, India perceived this new sale as tilting the balance in aircraft technology toward Pakistan. The combat effectiveness of the F-16 was assessed by Indian officials as the equivalent of three to four MiG-21s or three Jaguars.[57] In February 1982, India countered these moves by concluding a deal with France for the purchase and later manufacture in India of the advanced Mirage-2000, an aircraft that had not gone much beyond the test stage in France at the time of decision.[58] By 1985, seven Mirage-2000 aircraft had been delivered and another thirty-three were on order.

The seesaw arms race in front-line combat aircraft between India and Pakistan has been accompanied by Indian efforts to improve the quality of other aircraft in the air force. Thus, the earlier rejection of the MiG-23 has since been reversed; it is now being designated as the successor to the manufacture of MiG-21s in India. Meanwhile, the Indira Gandhi government also chose to acquire several Soviet MiG-25 reconnaissance-interceptor aircraft and the Soviet AN-32 transport aircraft to meet the needs of its aging interceptor and transport squadrons. More recently, still fearful and dissatisfied with these moves to offset the delivery of the forty American F-16s to Pakistan that began in 1984, India decided to purchase forty advanced Soviet MiG-29s immediately to provide an instant counterbalance to Pakistan's F-16s.[59] (As of early 1986, no MiG-29s had yet been delivered.)

It should be apparent from this discussion that the air force defense program was geared primarily to developments in Pakistan. There was little effort to counter the large but antiquated Chinese

[56] *New York Times*, June 16, 1981. Prime Minister Indira Gandhi's criticism of the American F-16 sale to Pakistan and India's decision to negotiate with France for the purchase of the Mirage-2000 are reported in the *New York Times*, July 11, 1981.

[57] *India News*, July 20, 1981. See also Maj. Gen D. K. Palit, "Mirage Deal Needs Careful Thought," *Times of India*, October 30, 1981.

[58] *Times of India*, February 13, 1982. For the present status of Mirage-2000 deliveries to India, see *The Military Balance, 1985–86*.

[59] *New York Times*, September 25, 1984. See also *The Military Balance, 1985–86*

air force of over five thousand combat aircraft. Since the war of 1962 did not involve the aircraft of either side except for transport and supply missions, most IAF planning tended to downgrade the possibility of a Chinese air attack. However, there is some recognition in India that China's technological problems of the 1960s have since been overcome. The ability of the IAF to meet a Chinese air attack rests mainly on its large fleet of interceptors and several Soviet SAM systems. Unlike defense programs to meet the Pakistani threat, there is no effort to produce a counterstrike air force capability that could hit targets deep in Chinese territory.

IAF defense programs in the mid-1980s are still linked to traditional security perceptions. The prospective modernization of the Chinese air force in the future with western assistance, the Pakistani acquisition of American F-16s, and the potential, though unlikely, Pakistani access to Mirage aircraft via some of the Islamic states of the Middle East are factors that may be considered in future IAF planning. Indeed, the increasing reach of the IAF Jaguars and the procurement of Mirage-2000s and MiG-29s are indications that air power beyond the subcontinent is already being taken into account in India's contingency planning.

In sum, IAF plans and programs of the 1980s have envisaged phasing out the Canberras, Hunters, Mystères, Gnats/Ajeets, Maruts, Sukhoi-7Bs, MiG-21s, and virtually all the transport aircraft that were procured more than two decades ago. These have been steadily replaced by MiG-23Bs for ground attack, MiG-25s for reconnaissance and interception, and Jaguars and eventually Mirage-2000s and MiG-29s for attack and strategic bombing. At least until the end of the decade, the MiG-21*Bis* will continue to provide air defense.[60] Meanwhile, India's own tactical strike aircraft developed at HAL is expected to move to the production stage by the early 1990s. For the GTX engine, a high thrust gas turbine engine capable of greater than Mach 2 performance, has been developed by HAL in Bangalore, and plans were being made in the mid-1980s to acquire a suitable airframe through indigenous design and production or through overseas purchase.

Force Levels and Defense Programs of the Navy

The growth of Indian naval forces and defense programs has been somewhat different from the other two services. The general neglect of the navy under both the British and the new government of

[60] *Times of India*, July 14, 1980. See also Jerrold F. Elkin, "The Indian Air Force of the 1980s," *Air University Review* 35, no. 6 (September-October 1984):88–95.

India for more than fifteen years after Independence meant any naval buildup would begin with an almost clean slate.[61] Although during the 1950s the navy was able to purchase from Britain an aircraft carrier, two cruisers, six destroyers of the "R" and "Hunt" class, and several antisubmarine and antiaircraft frigates, these were mainly overly used and dispensable vessels of the Royal Navy. Consequently, the modest growth of the Indian navy from the mid-1960s could be planned with reference both to the navy's newer combat needs and the prevailing international political constraints on weapons procurement. The result has been a predominantly Soviet-built Indian navy, unlike the relatively more mixed bag of western and Soviet equipment found in the other two services.

Just before the outbreak of the 1971 war, the navy continued to deploy the aircraft carrier *Vikrant*, two cruisers, three destroyers, and eight antisubmarine and antiaircraft frigates, all purchased from Britain twelve to fifteen years earlier.[62] However, the effects of the Indian decision to shift toward Moscow were already becoming evident. In 1971, deliveries had been made of four F-class submarines, five Petya-class frigates, and an unknown number of Osa-class missile boats. By 1985, dependence on Soviet vessels had become overwhelming. They now included three Kashin-2 destroyers with Styx SSMs (surface to surface missiles), eight F-class submarines, ten Petya-2 frigates, fourteen Osa-1 and -2 missile boats with Styx SSMs, three Nanuchka corvettes, five Poluchat large patrol crafts, six Natya ocean minesweepers, and seven Polnocny landing craft.[63] Also on order were two Kashin destroyers, four F-class submarines, two Nanuchka corvettes, and five more Polnocny landing crafts. What remained of the older British vessels were the aircraft carrier, one cruiser, and some frigates. However, a whole new line of Leander-class frigates was manufactured in the meantime by Mazagon Docks in Bombay in collaboration with Vickers and Yarrow of Britain. Six such frigates were delivered by 1983.[64]

There was also a modest growth in the naval air force, which in contrast to the main part of the Indian navy, was almost exclusively non-Soviet in composition. Its fifty aircraft and helicopters have consisted of one squadron of fifteen older British Sea Hawks of which ten have normally been stationed on the *Vikrant*. The helicop-

[61] See Thomas, "The Indian Navy in the Seventies."

[62] From *The Military Balance, 1971–72*.

[63] From *The Military Balance, 1985–86*.

[64] From *The Military Balance, 1982–83*. With the completion of the licensing period with Vickers and Yarrow of Great Britain, the Leander-class frigates have been renamed the Godavari-class frigates.

ter squadrons have consisted of one ASW (anti-submarine warfare) squadron of five Alizes, five ASW squadrons of ten Seakings and five Ka-25s, and one general purpose squadron of eleven Alouette-3s made in India under license from France. The major procurement in the 1980s to replace the Sea Hawks has been a purchase from Great Britain of VSTOL Sea Harriers. By 1985, eight Harriers had been delivered and ten more were on order. The only Soviet aircraft in the naval air wing were three Ilyushin-38s for maritime reconnaissance.

It should be apparent that the size of the Indian naval force as of the mid-1980s is still comparatively small for a country with a 3,400-mile coastline and a growing seaborne commerce. In the past this was partly due to the fact that there had been no major development of the Pakistan navy. Since then this situation has modestly changed. In 1985 the Pakistan navy consisted of eight destroyers, two Agosta and four Daphne submarines, another five SX-404 midget submarines; sixteen Chinese fast-attack craft, including twelve Shanghai-2 and four Hu Chwan fast-attack torpedo and gunboats; a few ASWS and general purpose helicopters, and some minesweepers. Three type-21 frigates and sixteen Harpoon SSMS were on order. A naval air wing was also being developed with maritime/ASW squadrons equipped with Exocet ASW missiles and sixteen helicopters for attack and antisubmarine warfare.

Although the Pakistani naval threat was never substantial, naval threats have been perceived previously from Indonesia under Sukarno and from Iran under the Shah, whose states had substantial numbers of imposing combat vessels. However, since the fall of both leaders, Indonesian and Iranian naval forces have declined and show no tendency of further development that would seriously threaten the Indian navy. In spite of these favorable conditions, the potential for growth by the Pakistani, Indonesian, and Iranian navies is still perceived as considerable, especially because all three states possess either substantial foreign exchange reserves or have access to such income. All three states also have the capacity to train efficient and capable naval manpower.

One significant aspect of naval development is the wide gap between the navy's envisaged role in the defense of India and actual naval defense programs sanctioned by the government. Nevertheless, the gap is much smaller than it was in the 1960s and naval hopes remain that it will eventually be closed. Unrealized naval plans include the setting up of a Southern fleet with headquarters at Cochin in addition to the present Western and Eastern fleets with headquarters at Bombay and Visakhapatnam. Each of the three fleet headquarters would be expected to carry its own complement of destroyers, cruisers, frigates, missile boats, and various other noncombat vessels. Suggestions have also been made that each of the three fleets should have an aircraft carrier as its flagship. Accompanying such an expansion would be a substantial increase in shipbuilding and dockyard facilities that would largely be a joint commercial-military program.

A major problem with proposed naval defense programs is that they tend to involve heavy capital investments and to require virtually custom-built ships, since the quantities needed are not substantial. This is unlike the production of tanks, aircraft, or transport vehicles, for which production lines can be established. Even in regard to the more optimistic expansion plans that have been proposed, the quantities do not provide an economic rationale for setting up a large shipbuilding base. The justification usually rests instead on the need for defense autarky and the possibility of gaining economic benefits if military and commercial shipbuilding are combined.

Although the more grandiose naval objectives are far from being realized, there have been moves toward more pragmatic goals in newer proposals. Although most of the major vessels have continued to be imported directly from the Soviet Union, a new line of Indian-designed frigates—the Godavari-class—is planned for production by Mazagon Docks in the 1980s to succeed the six Leander-class frigates. This new, indigenously designed frigate is expected to have an ASW system and to be able to carry two helicopters.[65] Less than 30 percent of the new line will be made of foreign components. Two of these frigates were delivered by 1985; two more were on order. Meanwhile, plans were finalized in 1981 for the licensed manufacture by Mazagon Docks of a line of West German SSK-1500 submarines with a 1,000 to 1,400 metric ton displacement.[66] Other naval defense programs have involved small craft production and dockyard expansion projects for the repair and modernization of existing vessels. These programs have all been conducted indigenously.

Defense-Related Nuclear Energy and Space Programs

Programs pursued under the Departments of Space and Atomic Energy are not classified in India as defense-related programs. Officially, all such programs are intended for peaceful purposes. As

[65] *Statesman*, December 8, 1979; *The Muslim* (Islamabad), March 10, 1980.

[66] *Tribune*, April 29, 1980.

early as the 1950s, the role of nuclear energy to meet the long-term energy needs of the industrial sector was stressed by the government under Prime Minister Nehru. The international oil crisis since 1973 has only reinforced the Indian commitment to nuclear energy to fill the critical gap between future industrial demands and anticipated shortfalls in fossil, thermal, and hydroelectric sources of energy.[67]

Nevertheless, a major issue underlying the Indian commitment to nuclear energy is New Delhi's interest in nuclear weapons. The peaceful nuclear program must also be seen in the context of India's space exploration program, which implicitly amounts to the growth of Indian ballistic missile delivery capabilities. Consequently, the military intent cannot easily be ignored. Since these programs do not fall under the jurisdiction of the Ministry of Defence and because their weapons potential cannot be easily separated from their peaceful civilian purposes, the programs of the Departments of Atomic Energy and Space need to be assessed as a whole.

Allegations of military motives underlying India's nuclear energy development stem mainly from the questionable commercial viability of nuclear power as compared to coal-fired thermal power and hydroelectric power available in India. Cost estimates in the United States, West Germany, Japan, France, and other western nations suggest that the early optimism placed in nuclear energy from light water and heavy water reactors and later from fast breeder reactors now appears to be misconceived. The scarcity and high cost of uranium sources, the high capital cost and growing technical complexities in nuclear plant design and operation, and fears of environmental hazards are all said to have reduced the prospects for commercial nuclear energy in the future.

The pessimism found among certain political, economic, and technical circles in the West is not shared in India by the political leadership and the scientists and engineers in the Departments of Atomic Energy and Space. Indian nuclear scientists and engineers have argued that coal-generated electricity is not necessarily cheaper than nuclear-generated electricity. According to Raja Ramana, secretary of the Department of Atomic Energy, the apparent higher cost of nuclear power over coal power is due to the nature of the cost accounting that has been applied to these two sources of energy.[68] In a nuclear program, capital investment is needed in certain

[67] See especially *Estimates Committee, 1977–78*, 16th Report, Ministry of Energy, 6th Lok Sabha, pp. 216–41.

[68] See Raja Ramana, "Inevitability of Atomic Energy in India's Power Pro-

TABLE 5.2
Comparative Costs of Power

Coal-Generated Electricity		Nuclear-Generated Electricity	
Cost Component	*Investment (Rs./KWe)*[a]	*Cost Component*	*Investment (Rs./KWe)*
Power station	4,500	Power station	5,000
Coal mining	750	Heavy water plants	495
Coal transportation	1,000	Uranium exploration	65
		Uranium mining	295
		Fuel fabrication	185
		Fuel reprocessing	135
Totals	6,250		6,175

SOURCE: Raja Ramana, "Inevitability of Atomic Energy in India's Power Programme," in *Energy Policy for India*, ed Rajendra K Pachauri (New Delhi, Macmillan Co of India, 1980), p. 235.
[a] Rupees per kilowatt of electricity generated.

associated sectors, such as production of heavy water and the setting up of various projects to establish control over the nuclear fuel cycle. The entire capital outlay here has been included in the assessment of the average cost of nuclear-generated electricity. If similar accounting were applied to coal-generated electricity, different comparative costs of power (based on 1977 figures) would appear (see table 5.2).[69]

Although there are drawbacks to pursuing a nuclear energy program in India, Indian political leaders and analysts point out that there are drawbacks in the generation of conventional electricity as well. In the case of coal, problems arise from its location and in mining and transporting it. Coal mines are concentrated in the northeast of India and distribution throughout India is both expensive and may be paralyzed by a nationwide strike by railway workers. Similarly, the few rivers and their uneven location in India means that this source of energy is limited to certain regions. Although these two sources of energy will continue to generate the bulk of electricity in India for several decades to come, the growth in ther-

gramme," in *Energy Policy for India*, ed. Rajendra K. Pachauri (New Delhi: Macmillan Co. of India, 1980), pp. 221–38.

[69] Raja Ramana, "Inevitability of Atomic Energy," p. 235. Another official discussion demonstrating the comparative advantage of nuclear power over coal-generated power is found in the brief section on the "economics of nuclear power," in *Performance Budget of the Department of Atomic Energy, 1980–81* (New Delhi: Department of Atomic Energy, Government of India), pp 3–5

mal and hydroelectric generating capacity is not expected to keep pace with industrial and urban demand. Nuclear energy is seen as a means of filling the projected gap. Thus, despite the relatively small consumption of nuclear energy in India, it is perceived to play a critical role in India's development strategy.

Another major argument suggesting a peaceful intent in India's nuclear energy development is the poor economic trade-off between a nuclear weapons program and the present conventional Indian defense program. In an economy with an abundance of cheap labor, a highly capital-intensive nuclear weapons program seems less desirable than the present emphasis on the labor-intensive army defense program and even air force and naval defense programs.[70] Nevertheless, whether intended or not, India's nuclear and space programs carry military benefits.

Whatever the economic justification—or lack thereof—for India's nuclear weapons program, there has been considerable growth in this direction. (See table 5.3.) In 1981, India had nuclear power stations at four locations in the states of Maharashtra (Tarapur, Bombay), Rajasthan (Kota), Tamil Nadu (Kalpakam, Madras), and Uttar Pradesh (Narora).[71] The Tarapur station, with a capacity of 420 MWe (megawatts of electricity) was the first to be commissioned in 1969. The station consisted of two General Electric light water reactors of 210 MWe each, furnished on a turnkey basis by the United States with about 30 percent of the costs paid by Indian industry and the same percentage of participation by Indian labor. The Rajasthan power station carried a capability of 440 MWe and consisted of two heavy water units supplied by the Atomic Energy Commission of Canada. This station began operation in 1973 and was built with 70 percent of participation and costs provided by Indian industry and labor. Since then, two heavy water reactors with a total capacity of 470 MWe were installed at Kalpakkam near Madras while two similar reactors with the same capacity were being built at Narora in Uttar Pradesh. When the Kalpakkam and Narora units are fully commissioned in the mid- to late 1980s, India is expected to have a total nuclear electricity generating capacity of 1,800 MWe. Meanwhile, ten Narora-type 235-MWe reactors are planned to be completed by the early 1990s, and twelve more nuclear reactors carrying 500 MWe capacity each are planned to be completed by the year

[70] For the rationales underlying the distribution of the Indian defense budget, see Thomas, "The Armed Services and the Indian Defense Budget."

[71] See *Nuclear News* 24, no. 2 (February 1981); and *Estimates Committee, 1977–78*, 16th Report.

TABLE 5.3
India's Nuclear Energy Program

a. Installed Electricity Generating Capacity, 1979 (MWe)[a]

Hydro	*Thermal*	*Nuclear*	*Total*
11,610 (38%)	16,450 (58%)	1,095 (2%)	29,163 (98%)[b]

b. Projections of Nuclear Power Generating Capacity as of 1979, with updating

Year	*MWe*	*Status*
by 1981	1,095	achieved in 1985
by 1991	8,620	likely achievement: 5,000 MWe
by 2000	10,000	likely achievement: 8,000 MWe

c. Existing Nuclear Plants (installed or being completed as of 1985)

Location	*Power Plants*	*Type of Reactor*	*MWe*	*Status*
Tarapur, Bombay	TAPP-I & II	light water	420	operational 1969
Kota, Rajasthan	RAPP-I & II	heavy water	440	operational (I-1973; II-1981)
Kalpakkam, Madras	MAPP-I & II	heavy water	470	partially operational (I-1985)
Narora, U.P.	NAPP-I & II	heavy water	470	not yet operational
Total Capacity			1,800	

d. Future Nuclear Plants Planned (all indigenous reactors)

Year	*No. and Description*	*MWe*
by 1991	10 Narora-type	235 ea. (2,350 total)
by 2000	12 new standardized design	500 ea. (6,000 total)

e. Heavy Water Plants (installed capacities but not necessarily operational as of 1985)

Location	*Metric Tons/Year*	*Status*
Nangal, Punjab	14	operational 1962
Baroda, Gujarat	67	operational 1977
Tuticorn, Tamil Nadu	71	operational 1978
Talcher, Andhra Pradesh	62	being commissioned
Kota, Rajasthan	110	being commissioned

f. Nuclear Research Reactors

Plant	*Capacity*	*Year Operational*
Apsara	400 KW	1956
Cirus	40 MW	1960
Zerlina	100 W	1961
Purnima	0 KW	1972
14 MW fast breeder research reactor	14 MW	1985
100 MW thermal research reactor (Dhruva)	100 MW	1985

g. Plutonium Extraction Capacity

Plant	*Kg/Year*
Tarapur	100
Kalpakkam	100

TABLE 5.3 (*cont.*)

Cirus Research Reactor	40
Dhruva	100

SOURCE Compiled by the author from the annual and ad hoc reports of India's Department of Atomic Energy, Ministry of Energy, Ministry of Information and Broadcasting, and from Indian and American newspaper reports

[a] Megawatts of electricity generated.

[b] Oil-fired plants not included.

2000. Some of these new reactors are expected to be fast breeder reactors. Such plans, if successful, would put 10,000 MWe of nuclear electricity in the power grids of India by the turn of the century. Almost all of this will have been developed through the exclusive participation of Indian industry and labor. These goals are expected to be realized in a three-stage process, moving from natural uranium-fueled heavy water reactors to fast breeder reactors, and finally to the employment of primarily thorium-fueled fast breeder reactors.[72] Since India has vast resources of thorium, the nuclear energy program is ultimately expected to become both indigenous and commercially viable in the next century.

The fast breeder reactors, when commissioned, will generate considerable weapons-quality plutonium. India's nuclear energy objectives thus have considerable significance for Indian nuclear weapons capability. In fact, even at present, the Tarapur and Kalpakkam nuclear reactors carry reprocessing capacities of spent uranium fuel that would enable India to generate about 880 pounds of weapons-grade plutonium annually—enough to keep pace with the growth of Chinese nuclear weapons capability.[73]

Similar but less forceful arguments are advanced in the promotion of the space program. (See table 5.4.) This program is said to advance Indian development in the areas of meteorology, satellite communications for television and radio broadcasting, and satellite observation of the earth in order to aid geological exploration of natural resources. But the parallel strides of India's space program alongside those of the nuclear energy program may camouflage a nuclear weapons delivery capability.

The Indian Space Research Organisation is one of the country's

[72] See *Estimates Committee, 1977–78*, pp. 218–19; and *Performance Budget of the Department of Atomic Energy, 1980–81*, Dept. of Atomic Energy, Government of India, pp. 3–4.

[73] Report by *Los Angeles Times* correspondent Tyler Marshal, reprinted in the *Boston Globe*, March 1, 1981.

TABLE 5.4
Satellites and Launch Vehicles

Satellite	*Date of Launch*	*Launch Vehicle*
Aryabhata (360 kg) (Ind. design & make)	April 1975	Soviet rocket carrier
Bhaskara-I (444 kg) (Ind. design & make)	June 1979	Soviet Intercosmos spacecraft
Rohini-1 (35 kg) (Ind. design & make)	July 1980	Indian-made rocket, SLV-3
Rohini D-1 (41 kg) (Ind. design & make)	May 1981	Indian-made rocket SLV-3(D)-1
APPLE (670 kg) (Ind. design & make)	June 1981	European Space Agency's Ariane rocket
Bhaskara-II (447 kg) (Ind. design & make)	November 1981	Soviet Intercosmos spacecraft
INSAT-IA (1,050 kg) (Ind. design; U.S. make)	April 1982	United States Delta rocket carrier
INSAT-IB (1,193 kg) (Ind. design; U.S. make)	August 1983	United States space shuttle Challenger

Planned Future Satellites and Launch Vehicles (1985–1990)
(All the projects noted here are indigenous.)

Satellites

Indian Remote Satellite (IRS-1), approx. 600 kg
Stretched Rohini Satellite Series (SROSS)
INSAT-II series and Proto-INSAT

Launch Vehicles

Augmented Satellite Launch Vehicle (ASLV): payload, 150 kg; altitude, 400 km
Polar Satellite Launch Vehicle (PSLV): payload, 600 kg; altitude, 500–1000 km
Vikas rocket: payload, 600 kg; altitude 900 km

largest and most successful groups of scientists and technocrats. Its main center is at Thumba in Kerala, with other smaller centers at Ahmedabad in Gujarat, Bangalore in Karnataka, and Sriharikota in Rajasthan. After successfully launching several atmospheric satellites in experiments for weather forecasting, India in July 1980 launched the indigenously built four-stage rocket *SLV-3* carrying the *Rohini-1* satellite.[74] The rocket was 22 meters long and weighed 17 metric tons; the satellite weighed 35 kg. *Rohini-1* was India's third satellite launched; the other two, *Aryabhata* and *Bhaskara*, were launched from the Soviet Union with Soviet rockets.

For the decade 1980–1990, the government has tentatively allot-

[74] *Military Yearbook, 1980–81* (New Delhi: Guide Publications, 1980), pp. 17–26.

ted Rs. 800 crores ($1 billion) for the space program, an amount that is likely to be revised upward substantially in the years to come. As many as twenty-three satellites, including three major ones with payloads in the 450–600 kg range, are planned for launching during the next ten years.[75] All the launches will carry technological and scientific payloads heavier than the one in July 1980. A major demonstration of India's rocket capability is expected in 1986 or 1987 when an indigenously built 600-kg Indian Remote Sensing Satellite, the *IRS-1*, will be launched aboard a heavy Polar Satellite launch vehicle incorporating the latest propellant technology.

Even anticipating periodic setbacks, these nuclear and space programs could give India a powerful and credible military nuclear posture in the years ahead, and the roles of the Departments of Atomic Energy and Space in the defense of India cannot be understated. In effect what is occurring are reinforcing rationales. Whether or not these programs are desirable in terms of alternatives available or development priorities, they provide military opportunities. Subsequently, arguments revolving around the urgency of defense needs provide the case for continuing the programs regardless of whether they are economically justifiable from a civilian standpoint. This approach has produced serious debate in and outside the policy-making establishment, especially when there is a conflict between military and civilian priorities. The Departments of Atomic Energy and Space are aware of their dual civilian-military roles and their programs are accordingly formulated and executed.

III. The Defense Budgetary Process

Competing claims for India's scarce economic resources require that the defense sector convincingly justify its programs against proposed civilian programs that may provide direct economic benefits to the population. Although external threat conditions may seem like the main arbiter of Indian defense budgets, domestic political and economic conditions play an equally important part. To a considerable degree, the size and specifics of the defense budgets depend on the internal political strengths of various actors, both civilian and military, who are demanding a larger share of the national budget. When the Ministry of Finance is considering alloca-

[75] See *Hindustan Times*, December 15 and 30, 1980; and *Times of India*, December 15, 1980.

tions to various sectors of the economy, the claims of the Ministry of Defence must be weighed against the claims of other ministries such as Food and Agriculture, Industries, Commerce, and Health and Family Welfare.

Defense, Finance, and Budgetary Allocations

Much of defense policy in India is constrained by certain routine financial and budgetary procedures as well as the availability of economic resources, especially foreign exchange.[76] The latter dependency makes the relationship of the Defence Ministry to the Finance Ministry a critical one. In one respect this is not unusual; all other ministries bear this relationship to the Ministry of Finance to varying degrees. An increase in allocations over the previous year requires sanction at the cabinet level. Ultimately, it is the Ministry of Finance that acts as the clearing house for various demands on scarce resources. No doubt the cabinet determines priorities based on ideological beliefs, political necessities, election promises, and various demands of private economic lobbies, communal groups, and social classes. But the finance minister holds a pivotal position in persuading the cabinet what these priorities ought to be; he is able to emphasize his own personal ideological bent and policy preferences as well as the need to balance the budget and maintain the stability of the economy.

Deciding "how much" should go "where" is a challenging proposition. Establishing economic criteria for the distribution of budgetary resources among the ministries and departments can be a difficult and often arbitrary task. Maximum rates of economic growth may be a strong objective, yet even this standard necessarily has political overtones and consequences; it will favor certain classes, regions, or groups of people over others. Prevailing political necessities and the ruling party's ideological outlook are therefore crucial in determining planning priorities, and these in turn will indicate how varying amounts should be distributed. Parliamentary debates provide a forum for ascertaining and solving the distribution questions.

In the case of defense, however, deciding "how much is enough" through parliamentary debate is complicated by the general lack of knowledge among members on matters of defense and strategy. The debate in the Lok Sabha on the "Demands for Grants" of the

[76] Parts of this section are adapted from Thomas, *Defence of India*, pp. 89–96.

Ministry of Defence is usually of a broad and general nature and rarely deals with the specifics of budgetary requests by the services.

Determining the size of the defense budget consequently rests largely with the Ministries of Finance and Defence, and placing constraints on demands by the Defence Ministry and the services is then primarily done by the Ministry of Finance. In order to avoid major last-minute confrontations and rejections when the budget is considered in Parliament, it is the practice of the Ministry of Finance to designate an advisor who works with officials of the Defence Ministry on a year-round basis.[77] The advisor ensures that proposals are scrutinized during the formative stages in order that policy proposed is not unreasonable or difficult to implement from a budgetary standpoint. The advisor heads a special wing of the Finance Ministry that deals solely with defense expenditures. His subdivision is expected to work in close collaboration with the Defence Ministry and the armed services' headquarters but to retain its independence. The functions of the Ministry of Finance (defense wing) is therefore of both an advisory and a controlling nature, its objective officially defined as the "safeguarding of economic efficiency and propriety in public defense."[78]

Thus once overall policy has been established at the committee level, the size and direction of defense expenditures is strongly influenced by the Ministry of Finance. The finance minister has only to say that no foreign exchange can be put aside for a proposed defense project in order to kill it.[79] Even when foreign exchange may not be critical, strong economic reasons may be put forward for directing resources elsewhere rather than to defense. As evident in the broader concept of security in India, a collapsing economy could be just as much a danger to national security as low spending on defense production and the armed services.

The relationship between Finance and Defence, therefore, should be a key factor in determining the size of Indian defense budgets. The importance of this relationship has been enhanced in part by the existence of weak political opposition in Parliament until 1977, which could provide neither alternative policies nor serious criticism of budgets for various defense programs. Since no other party or coalition could hope to replace the Congress party at the

[77] This procedure is described in the appendix of the annual *Defence Services Estimates* (New Delhi: Ministry of Defence, Government of India).

[78] *Defense Services Estimates, 1970–71*, appendix.

[79] Michael Brecher, *India and World Politics: Krishna Menon's View of the World* (London: Oxford University Press, 1968), p. 240.

TABLE 5.5
Annual Defense Expenditures

Year	Total Defense Expenditure (Rs. crores)	Total Government Expenditure (Rs. crores)	GNP (Rs crores)	Defense Expenditure as % of Total Govt. Expenditure	Defense Expenditure as % of GNP
1964–65	806	2,603	21,113	31.0	3.8
1965–66	885	2,720	21,866	32.5	4.0
1966–67	909	3,217	25,250	28.3	3.6
1967–68	968	3,148	29,612	30.7	3.3
1968–69	1,033	3,140	30,293	32.9	3.4
1969–70	1,101	3,590	33,521	30.7	3.3
1970–71	1,199	4,120	36,452	29.1	3.3
1971–72	1,525	5,498	38,972	27.7	3.9
1972–73	1,652	5,498	42,939	30.0	3.8
1973–74	1,681	5,845	53,447	28.8	3.1
1974–75	2,112	74,423	62,972	28.5	3.4
1975–76	2,472	9,429	66,193	26.2	3.7
1976–77	2,563	10,291	71,826	24.9	3.6
1977–78	2,634	11,605	81,105	22.7	3.2
1978–79	2,868	13,366	86,927	21.5	3.3
1979–80	3,356	14,473	96,850	23.2	3.5
1980–81 (RE)	3,800	18,428	n.a.	20.6	n.a.
1981–82 (BE)	4,200	19,385	n.a.	21.7	n.a.

SOURCE: *India, 1981* (New Delhi: Publications Division, Ministry of Information and Broadcasting, 1981), p. 34

NOTE: RE–revised estimates, BE–budget estimates, n.a.–not available; 1 crore = 10 million

federal level at the time, opposition views and alternatives could be safely brushed aside.

In practice, this situation did not lead to extreme demands by the Defence Ministry but, strangely, to moderation, even during the years immediately after the Sino-Indian war of 1962. Defense spending formed an unprecedentedly high 28 percent of total central government expenditures in 1963–1964, the first planned defense after the 1962 Sino-Indian war. (See tables 5.5 and 5.6.) Between 1964 and 1976, defense expenditures as a percentage of central government expenditures remained high, between 28 and 33 percent. This compares with an average of 15–18 percent (1.9 percent of GNP) between 1950 and 1962 and about 21–25 percent (3.5 percent of GNP) between 1976 and 1982. But the relationships between defense and finance ministers have varied. Such variations

TABLE 5.6
Ministry of Defence Budget Estimates for 1980–1981
as Presented to Parliament, 1980

	Estimates (millions of Rs)	*% of Total and % within Category*
Ministry of Defence	2,148	5.8
DEFENCE SERVICES	30,353	81.7
Army	19,933	65.7
Air Force	8,276	27.3
Navy	2,144	7.0
DEFENCE CAPITAL OUTLAY	2,641	7.1
Army	739	28.0
Air Force	260	9.8
Navy	1,046	39.6
Ordnance Factories	507	19.2
Research & Development	64	2.4
Inspection of Works	25	1.0
PENSIONS	2,012	5.4
Army	1,822	90.6
Air Force	122	6.0
Navy	68	3.4
Totals	37,154	100.0

SOURCE *Demands for Grants of the Ministry of Defence, 1980–81*, Lok Sabha, Parliament House, New Delhi, p. 3.

NOTE The Defence Services cover the cost and maintenance of military personnel; Defence Capital Outlay covers plant and equipment costs.

have depended on the importance attributed by the prime minister to the defense and finance portfolios within the cabinet, the personalities who have held these posts, and the prevailing political circumstances, both domestic and international. Essentially, the defense-finance relationship has had four discernible phases: the pre-Krishna Menon era from 1947 to 1957, the Krishna Menon-Morarji Desai phase from 1957 to 1962, the post-Menon era from 1962 to 1977, and the phase following the first defeat of the Congress party at the national elections in 1977.

The first three defense ministers of independent India—Bladev Singh, Gopalaswami Ayyangar, and Kailash Nath Katju—played a relatively minor part in both policy making and in ascertaining the size and shape of defense budgets. The finance ministers of the time—R. K. Shanmugam Chetty, John Mathai, and C. D. Deshmukh—were immersed in the internal policies being drawn up for the economic advancement and political consolidation of the new republic. These men were overshadowed by the figures of Jawaharlal Nehru, Vallabhbhai Patel, Govind Ballabh Pant, and Maulana Azad. The major external situation of the period that called for any semblance of military policy was the Kashmir issue. Policy on this was first a compromise between the conflicting outlooks of moderation and force represented by Nehru alone.[80] Meanwhile, nominal defense plans and budget programs were mechanically pushed through Parliament with little contribution by defense and finance ministers. The only serious foreign policy debates that occurred in Parliament were over the signing of the Sino-Indian treaty on Tibet and the Pakistani decision to enter the western SEATO and CENTO defense pacts.

Between 1957 and 1962, the portfolios of defense and finance were held by the left-wing Krishna Menon and the right-wing Morarji Desai. Although this was the first time that someone close to Nehru had held the defense post, there was little change in defense allocations. In the years between 1951–1952 and 1961–1962, defense spending never rose beyond 2 percent of the GNP despite the presence of Menon. In part this may have been due to his unpopularity with other cabinet members and in the rest of the Congress party and because of his alleged excessive interference with appointments and promotions in the armed services. The considerable reactivation of defense production that occurred under Menon and his interest in the affairs of the armed services occurred, therefore, within the limits of defense budgets that continued to reflect a governmental policy of complacence—or perhaps confidence—on national security matters.

The financing of defense programs before the Sino-Indian war and the resignation of Menon in 1962 was strongly influenced by the Ministry of Finance. Menon had argued that the failure to meet the Chinese threat in 1962 was primarily the fault of the Ministry of Finance, headed by Desai. With reference to the Planning Commission and the Finance Ministry, Menon once said: "Ultimately all Finance Ministers—especially Morarji Desai and T.T.K. [T. T. Krishnamachari, one of the Indian finance ministers]—used to say, 'That is all (the money) that I can give.' " T.T.K. used stronger language than others; Morarji was more dogmatic.[81]

[80] Michael Brecher, *Nehru: A Political Biography* (London: Oxford University Press, 1959), pp. 151–71.

[81] Brecher, *India and World Politics*, p. 271.

After the debacle in 1962 and the resignation of Menon, the relationship between defense and finance ministers entered a new phase. Y. B. Chavan served as defense minister from 1962 to 1966. Desai continued briefly as finance minister followed by a much weaker T. T. Krishnamachari, who had also held the post in the mid-fifties. Krishnamachari was followed for a year by Sachindra Chaudhuri and then again by Morarji Desai. The former defense minister Y. B. Chavan then took over the finance portfolio. Meanwhile, the former external affairs minister, Swaran Singh, took over as defense minister and was followed by a former food minister, Jagjivan Ram.

According to a former cabinet secretary, S. S. Khera, the advent of Chavan as defense minister in the midst of the shock of defeat in 1962 produced a more open and sympathetic relationship with the finance minister as well as with other cabinet ministers. Chavan's requests for defense funds "were better explained, better understood, and more readily met than those of his unfortunate predecessor, Krishna Menon."[82] Indeed, in the budget (fiscal) years 1963–1964 and 1964–1965, the first two years of Chavan's tenure as defense minister, defense allocations exceeded actual utilization by Rs. 51 crores and Rs. 48 crores respectively.[83] Only twice in later years did the Finance Ministry propose to Parliament an amount greater than the Defense Ministry could absorb (this occurred when Swaran Singh held the defense portfolio, and the amount never exceeded Rs. 10 crores).

The years after Chavan's departure from the Defence Ministry continued to produce a moderate-conservative relationship between defense and finance. The extreme case of Menon on the left and Desai on the right has not existed since the Sino-Indian war of 1962. However, whatever the relationship, there can be little doubt that finance ministers in India hold a near-veto position on the allocation of defense funds. The Finance Ministry's ability to squash defense programs at an early stage is also aided by the presence of its advisory "finance cell" within the Defence Ministry but responsible to the finance minister.

The rise of the Janata party to power in 1977 marked a new phase in the defense-finance relationship primarily because of the greater attention the government in power had to pay to the opposition in Parliament. The relationship was no longer the internal party affair

[82] S. S. Khera, *India's Defence Problems* (New Delhi: Orient Longman, 1968), p. 47.

[83] From the *Defence Services Estimates*, 1963–64 to 1972–73.

it had been for thirty years. Although relationships between Defence Minister Ram and Finance Minister Patel in the Janata government of Prime Minister Desai did not seem significantly different from that of the earlier Ram (defense) and Chavan (finance) relationship under Prime Minister Gandhi, defense programs were subjected to greater scrutiny in Parliament. In part this was because the opposition bench during the terms of both the Janata and later Congress governments has carried in its ranks former cabinet ministers, whose experience has made them better able to challenge ministry decisions.

Finally, a somewhat unique situation has prevailed since the return of the Congress government in 1980. Prime Minister Indira Gandhi held the defense portfolio concurrently until 1982; R. Venkataraman was finance minister. On the one hand, this relationship suggested the higher ranking and power of defense over finance and a greater attention to the needs of the services and their programs. On the other hand, it suggested the neglect of the defense sector because the prime minister was concerned with a variety of problems concerning many of the ministries. In retrospect, the defense sector appears to have gained during this period; major programs were initiated especially for the navy and the air force. Subsequently, faced perhaps with the pressures of the dual role, Mrs. Gandhi passed on the mantle of defense to her trusted finance minister, R. Venkataraman, thereby also signaling the high importance she continued to attach to defense matters. In September 1985, Mrs. Gandhi proposed R. Venkataraman as the next Vice President of India and he was duly elected to this prestigious but more ceremonial post a month before the prime minister's assassination.

In contrast to the interest and involvement of Mrs. Gandhi in defense matters, Prime Minister Rajiv Gandhi has—at least in the early phase of his term in office—paid less attention to external diplomatic and strategic affairs. Domestic political and economic issues have largely consumed his interest. Following Rajiv's victory in the December 1984 general elections, the defense portfolio was first briefly held by the former minister of external affairs, P. V. Narasimha Rao, and subsequently by the former minister of home affairs, S. B. Chavan.

The Armed Services and Budgetary Distributions

The strategic perspectives and defense programs of the services, discussed earlier, are to a certain degree reflected in the annual allocations to the Ministry of Defence. (See tables 5.7, 5.8, and 5.9.) The budget shares of the army, air force, and navy since the 1971 Indo-Pakistani war (table 5.8) indicates a steady percentage distribution of approximately 70 percent, 20 percent, and 10 percent, with gradual and marginal changes taking place toward the end of the decade. From 1978 on, the army's share of the service budget began to decline to about 63 percent; the share of the air force increased to about 27 percent. This compares with about 76 percent allocated to the army, 20 percent to the air force, and 4 percent to the navy during the five years after the 1962 Sino-Indian war (see table 5.9). Thus the navy's share rose by about 5 percent from the mid- to the late sixties and the air force's share rose by about 5 percent from the mid- to late seventies. Both times this increase occurred at the expense of the army's share.

A comparison of the budgetary distributions among the three services in the 1960s and 1970s demonstrates the perpetuation of two historical legacies. Up to the late 1960s, the allocations were still colored by the role and relative importance of the army under British imperial rule in India. The army was the dominant service, the air force had received some attention during the Second World War, and the navy had always been neglected. The government of India inherited this structure in 1947 and then perpetuated it. On the one hand, it was aided by the prevailing bureaucratic strengths within the military as well as by a certain degree of civilian inertia until 1962. On the other hand, the structure was justified by the nature of the military threat posed by Pakistan. The government of India had inherited one-third of the British Indian military service and apparatus and, too, had perpetuated the predominantly army-oriented defense apparatus. Thus the dominant threat faced by India was first by land and second by air. A threat at sea did not exist.

In the 1960s, the wars with China and Pakistan firmly established a dual threat. This called for a substantial overall increase in the Indian defense budgets but resulted in only marginal adjustments in the distribution pattern; the shares of the services remained approximately the same. Much of the increase allocations after 1962 went to the Himalayan borders with China and the raising of a second armored division to balance Pakistan's two armored divisions.

From the mid-1960s on, with no apparent change in the threats perceived as coming by sea, the navy began to gain attention. Its share rose from 4 percent in 1965 to almost 10 percent in 1973. During this time, the Pakistan navy continued to be neglected and

TABLE 5.7

The Indian Defence Budget, 1970–1980 (in millions of Rs.)

	1970–71	1971–72	1972–73	1973–74	1974–75	1975–76	1976–77	1977–78	1978–79	1979–80	1980–81
Service Revenue share[a]	11,030	13,901	14,958	15,461	19,711	23,043	24,213	25,247	26,896	30,145	30,352
Service Capital share[a]	1,170	1,400	1,718	1,523	1,475	1,709	1,623	1,996	2,142	2,048	2,043
Total Service budget	12,200	15,301	16,676	16,984	21,186	24,752	25,836	27,243	29,038	32,193	32,395
Total Defence budget[b]	12,969	16,204	17,653	18,102	22,569	26,381	27,563	29,257	31,336	33,872	34,543

SOURCE Figures derived from *Defence Services Estimates*, Ministry of Defence, Government of India. Figures for the years 1977–1978 to 1980–1981 are based on the budget estimates presented to the Lok Sabha. The rest are based on actual amounts spent.

NOTE Service Revenue category covers cost and maintenance of military personnel; Service Capital, the plant and equipment costs The Service budget total is the sum of Revenue and Capital; the Defence budget total is the sum of the Service budget and miscellaneous expenditures.

[a] Allocations to defense public sector undertakings and ordnance factories are not included in the Revenue and Capital allocations of the Service budget.

[b] Total Defence budget figures do not include pensions. Pensions usually constitute 5–6 percent of the Defence budget. In 1980–1981 the army absorbed 91 percent of the total pension allocation, the air force 6 percent, and the navy 3 percent.

TABLE 5.8

The Budget Share of the Armed Services, 1970–1980 (in percent)

	1970–71	1971–72	1972–73	1973–74	1974–75	1975–76	1976–77	1977–78	1978–79	1979–80	1980–81
ARMY											
Revenue allocation[a]	74	75	75	73	74	73	72	71	70	65	66
Capital allocation[a]	51	51	50	38	31	36	38	35	36	36	36
Service budget	72	73	72	70	71	70	70	68	68	63	64
Defence budget[b]	67	69	68	65	66	66	65	64	62	60	60
AIR FORCE											
Revenue allocation[a]	21	20	20	22	21	21	22	22	22	28	27
Capital allocation[a]	15	17	15	17	21	15	13	13	13	12	13
Service budget	20	20	20	21	21	21	21	21	22	22	21
Defence budget[b]	19	19	19	20	20	19	20	20	20	25	25
NAVY											
Revenue allocation[a]	5	5	5	5	5	6	6	7	8	7	7
Capital allocation[a]	34	32	35	45	48	49	49	52	51	51	51
Service budget	8	7	8	9	8	9	9	10	11	10	10
Defence budget[b]	8	7	7	8	8	8	8	9	10	9	9

SOURCE Figures derived from *Defence Services Estimates*, Ministry of Defence, Government of India. Figures for the years 1977–1978 to 1980–1981 are based on the budget estimates presented to the Lok Sabha. The rest are based on actual amounts spent.

NOTE *Figures may not add to 100% due to rounding. See also note to table 5.7*

[a] Allocations to Defence public sector undertakings and ordnance factories are not included in the Revenue and Capital allocations of the Service budget.

[b] Total Defence budget figures do not include pensions. Pensions usually constitute 5–6 percent of the Defence budget. In 1980–1981 the army absorbed 91 percent of the total pension allocation, the air force 6 percent, and the navy 3 percent.

TABLE 5.9
The Share of the Armed Services in the Annual Defence Budget, 1962–73

	1962–63	1963–64	1964–65	1965–66	1966–67	1967–68	1968–69	1969–70	1970–71	1971–72	1972–73[a]	1973–74[a]
Revenue Total (Rs. crores)	436.8	727.9	715.8	790.9	834.1	906.4	970.1	1013.5	1103 0	1390.1	1427.4	1447.4
Army (%)	77.5	78.4	78.6	77.6	77.9	78.4	77.6	75.6	74.0	75.1	75.3	71.1
Air Force (%)	17.8	18.5	18.0	18.6	17.9	17.3	18.2	19.8	20.7	19.9	19.6	22.9
Navy (%)	4.7	5.1	3.3	3.8	4.2	4.3	4.2	4.6	5.2	5.0	5.1	6.0
Capital Total[b] (Rs. crores)	40.6	88.0	73.8	70 4	71.0	76.4	78.7	97.4	117.0	139.9	160.0	147.8
Army (%)	45.7	53.4	58.1	53.9	55.4	56.4	52.0	52.9	51.1	50.0	45.2	38.5
Air Force (%)	41.0	39.8	34.6	38.4	36.8	31.0	25.0	18 2	14.6	17.0	15.2	12.7
Navy (%)	13.3	6.8	7.3	7.7	7.8	12.6	23.0	28.9	34.3	33.0	39.6	48.8
Revenue Capital, and Noneffective Total[c] (Rs. crores)	495.7	834.3	810.1	884.9	907.6	1008.8	1078.9	1145.5	1265.1	1580.8	1644.4	1655.6
Army (%)	73.6	76.2	77.2	76.2	76.1	77.3	76.3	74.3	72.7	73.5	72.9	69.0
Air Force (%)	19.1	20.3	19.1	19.7	19.4	17.9	18.2	19.1	19.5	19.2	18.6	21.3
Navy (%)	5.3	3.5	3.7	4.1	4.5	4.8	5.5	6.6	7.8	7.3	8.5	9.7
% Increase from Previous Year												
Army (%)		69.5	−1.6	7.8	5.7	9.4	5.5	3.5	8.0	26.5	3.1	−4.7
Air Force (%)		79.3	-8.8	12.6	1.1	2.8	8.7	11.2	12.7	22.7	−0.1	16.2
Navy (%)		10.0	3.2	22.2	12.3	19.7	21.4	27.0	31.5	16.7	18.6	17.7

SOURCE Figures collated from the annual *Defence Services Estimates*, 1962–63 to 1973–74, Ministry of Defence, Government of India, New Delhi. Table published in Raju G. C Thomas, *The Defence of India A Budgetary Perspective of Strategy and Politics* (New Delhi. Macmillan Company of India, 1978, Columbia, Mo.: South Asia Books, 1978), p 157.

NOTE Figures may not add to 100% due to rounding.

[a] Figures for 1972–1973 are not actuals but revised estimates; those for 1973–1974 are from budget estimates. Actuals were not available at the time these data were collected in 1974.

[b] Note that capital total excludes expenses on manufacturing and research establishments directly assigned to one of the three services. It also excludes "Outlay on Industrial and Other Organizations," the eight public sector undertakings under the Defence Ministry.

[c] Noneffective charges refer to pensions, awards, etc.

as a consequence suffered a crushing blow in 1971 when warfare between the two countries was extended to sea for the first time. Meanwhile, the size and growth of the Indian air force remained steady; the essential effort here was to matching the strength of the Pakistan air force, which had been modernized in the late 1950s and early 1960s under the U.S. Military Assistance Program. During this period no serious effort was made to counter a possible Chinese air thrust across the Himalayas due to the heavy economic burden this policy would entail, the dubious role the air force could play against Chinese military and industrial targets at great distances from the border, and the perception that China's air force was somewhat antiquated in qualitative striking power.

Thus the defense budgetary distributions in the 1960s continued to display the legacy of colonial strategy; only in the latter part of the decade did defense capabilities reflect the dual Sino-Pakistani threat. A similar problem of lag was evident through much of the 1970s as defense allocations continued to reflect the traditional strategic conditions described in chapter two. Although the newer strategic conditions described as the "extended" strategic framework were acknowledged by the Ministry of Defence in its annual reports from 1973 to 1978, such perceptions were not expressed in the budget.[84]

The time lag may be attributed in part to the focus on internal politics arising from the Emergency and later the fall of the Congress government. The reason may also be found in the inevitable delay that occurs between the perception of new threats by decision makers and the ability of governments to respond appropriately in times of peace. Given the speculative nature of threats perceived, translation into specific defense programs will always be difficult, particularly in democratic systems.

In effect, since Independence the budget shares of the services have changed only marginally—a 10 percent decrease for the army and a corresponding 5 percent increase for both the navy and the air force. The Army would doubtless discourage any further attempt to cut its share. On the other hand, in the newer strategic circumstances, a revised distribution in the proportion of 50 percent (army), 30 percent (air force), and 20 percent (navy) may be advisable, achieved through a steady increase in the overall defense allocation rather than through a cut in the army's budget. Arguments

[84] See Indian Defence Ministry, *Report 1974–75*, p. 2; *Report: 1975–76*, pp. 1–3; and *Report: 1977–78*, pp. 1–3.

for such a distribution also rest on the need to maintain a "military balance" among the three services. Overwhelming superiority carried by one service could lead to a biased view of the prevailing external threat and to distortions in defense planning and budget allocations.

CHAPTER SIX

Economic Issues Underlying Defense Planning and Production

DEFENSE planning takes place in the context of the larger economic milieu. How many and which resources may be allocated to defense depends on India's economic growth and stability. Broad-based growth, especially in key areas of the industrial sector such as aerospace, electronics, shipbuilding, and the automotive industry, are prerequisites for self-sufficiency in defense production. Similary, an internationally competitive export-oriented economy and a healthy annual trade balance are needed to generate sufficient foreign exchange for the import of advanced weapons from abroad. On the other hand, defense allocations may hamper economic growth and stability. Large-scale domestic weapons production may adversely affect the level of savings available for civilian investments. Substantial or expensive weapons imports may severely restrict the import of machinery and goods critical for development purposes. Defense spending that tends to aggravate price inflation may destabilize the economy and further complicate development plans. Finally, these adverse effects may in turn generate political instability, making defense planning more difficult.

Economic issues underlying defense planning in India are discussed here under four broad themes: (1) the nature of the defense economy in general and of defense planning in India in particular, and official policies regarding the roles of the public and private sectors in defense production; (2) the available and desirable level of investable savings that may be channeled into the defense sector; (3) the opportunity costs and inflationary effects of defense spending on the economy; and (4) the various Indian perspectives on the impact of defense spending on economic development.

1. Defense Planning

The Basis of Defense Planning

Resource allocations to Indian defense programs must be assessed in terms of the mobilization potential of the nation and the consequences of any large-scale diversion of resources away from the development program. The question of economic mobilization concerns the defense potential of a nation; the question of economic consequences concerns the alternative uses of such funds for civilian development programs and the possibility of inflationary or other harmful effects on the economy.

At a conceptual level, the mobilization potential of a nation is the difference between the total amount of resources available and the minimum subsistence on which human beings can survive. War economies have sought usually to maximize the resources that can be diverted to the defense effort. During the Second World War, the defense allocations of some of the major combatants rose to about 80 percent of their gross national products. Following the October 1973 Arab-Israeli war, defense spending in Israel in 1975 was over 35 percent of its GNP, and that of Egypt about 50 percent.[1]

However, economic and political limitations exist in determining the Indian resources available for defense. Although the GNP is often suggested as the best indication of total resources available, in a country like India this may be unsatisfactory, because large amounts of resources still remain untapped or idle. Additionally, a substantial portion of the Indian GNP is not recorded due to large-scale economic transactions on the black market that do not show up in official GNP estimates. One Defence Ministry official rightly pointed out some years ago that both the figures on total resources available in India and the Indian allocation to defense expressed as a percentage of the GNP do not reflect the situation accurately.[2] In fact, what India allocates to defense is perhaps smaller than the annual average of 3.5 percent that is usually estimated, because official GNP figures do not tell the whole story.

Trying to determine a subsistence level for the Indian nation also encounters several problems. The fact that a difference exists between "voluntary" and "forced" subsistence levels during times of both peace and war in itself provides room for controversy. Prolonged wars may force nations to survive at the bare minimum. In India, long periods of peace and the short wars that have occurred have rarely allowed the government to impose harsh economic measures on the population in order to divert resources for defense.

[1] From *The Military Balance, 1980–81* (London: International Institute for Strategic Studies, 1980), pp. 96–97.

[2] Interviews with author; the official interviewed requested anonymity.

Whatever the total economic resources available for mobilization, their diversion to defense is tempered by three conditions: the time available for war mobilization; the level of defense technology and skills that are available during a given period; and the psychological motivation of the population to undergo hardships for the defense effort. First, war potential can apply only to those resources that can be employed in defense. The time available therefore becomes important in converting resources currently utilized for civilian programs into resources that advance defense programs. As a result, the war potential of India (i.e., resources currently utilized in the civilian sector) can range from zero if the attack is sudden and unexpected to near-subsistence levels if the attack has been anticipated sufficiently in advance and if the ensuing war itself is a war of attrition. Note, however, that in the case of nuclear war there may be no such thing as war potential. What is available for defense or for counterattack is only those weapons that already exist.[3] Before mobilization can be attempted the war may well be over.

In the conventional setting, defense preparedness will depend on the nature of the war to be waged, the weapons needed for that purpose, and the skills needed to man the weapons. In the case of the Indian army, the war potential is considerable. This was evident after the Sino-Indian war, when ten mountain divisions were swiftly raised and several ordnance factories for the manufacture of light arms and ammunition were quickly set up. On the other hand, in the Indian air force and navy this potential would be much lower. Bombers and destroyers cannot be produced swiftly and pilots and seamen cannot be trained as quickly as infantrymen can. Long-term and perhaps even remote contingency planning is more critical in the case of these two services. Similarly, if a nuclear threat to India from China is already perceived to exist, India's ability to deter a nuclear attack or to avoid nuclear blackmail would be nonexistent were a sudden Sino-Indian confrontation to occur. If a deterioration in Sino-Indian relations were more gradual, then the present policy of maintaining India's nuclear weapons option by pursuing civilian development of nuclear energy and space programs would enable the government to divert or upgrade part of its civilian nuclear program to the production of nuclear weapons and delivery systems.

However, time alone may not be helpful if the level of technological skills and resources are insufficient to produce the weapons

[3] See Charles J. Hitch and Roland N. McKean, *The Economics of Defense in the Nuclear Age* (New York: Atheneum Press, 1973), pp. 65, 312–24.

needed. The ability of smaller nations to raise the necessary resources and technological skills to meet external threats is limited. Time may be relevant only to the extent that in due course India may acquire the know-how and necessary manufacturing facilities within the limitations of its overall and potential economic and manpower resources. Under these circumstances, the relationship of India's development planning to its defense planning is crucial.

The motivation to mobilize is the third factor that qualifies a nation's defense capability. The amount of material well-being a country's citizens are willing to sacrifice for the defense effort will depend on the objectives of the government's foreign policy, the nature and rigor of its security programs, and the cultural and psychological outlook of its people, especially in reference to the nation's military objectives and purposes. British doggedness during the Second World War against the Nazis and Vietnamese endurance during their seemingly endless war against the Japanese, French, and Americans are good examples. However, the wars of the Indian subcontinent have been of relatively short duration and it is difficult to assess how long hostilities could be sustained at the cost of economic sacrifice by the majority of the Indian population. Ultimately the severities of prolonged military combat will need to be weighed against the political objectives sought. Under peacetime conditions the ability of governments to impose severe economic hardships for the defense effort is even further limited.

In discussing these modifying factors on mobilization capacity, the country's GNP has been assumed as the basis for measurement. As mentioned above, however, such a measurement may be unsatisfactory in developing countries. If much of the nation's resources are not employed, the GNP will be relatively small and will not be a true indicator of the capacity to produce. Larger defense programs may be carried out simply by drawing on idle resources or by utilizing idle capacities in existing manufacturing establishments. The concept of maximum defense potential must therefore be evaluated in terms of the nation's GNP at a high level of employment. Moreover, this upper limit must be viewed as a moving one, primarily upwards—especially in a developing nation such as India—as new natural resources and technological methods are discovered, though sometimes downwards, as existing raw materials are depleted.

Economic issues are a major restraining influence on how much security can be afforded. Interpretations vary on the economic consequences of defense programs, but broadly speaking, there are two sets of beliefs, one adverse and the other favorable. The first school claims that a diversion of resources to defense retards economic growth and development, tends to be inflationary, and encourages attitudes that favor an undesirable military-industrial complex. The growth of the military-industrial complex in turn is alleged to generate foreign policies that are confrontational, calling for a continuing buildup of armaments and armed forces. According to the second school, defense programs tend to accelerate economic growth and development or tend to reverse recessionary trends in the economy. The perceived need to maintain a strong defense tends to encourage a stronger sense of nationalism that in turn provides greater motivation to save and invest in both the military and civilian sectors. The economic consequences of defense programs constitute an important part of the political debate in India, which is discussed in greater detail below.

Indian Defense Planning

Because development is one of India's primary tasks, defense programs need to be formulated in the overall context of economic planning. This was recognized immediately following the 1962 Sino-Indian war. After an emergency defense procurement program was introduced in the 1963–1964 defense budget, in 1964 the first five-year defense plan, valued at Rs. 5,000 crores, was formulated and presented to the Lok Sabha. War with Pakistan in 1965 further emphasized the need to synchronize defense with development and led to the formation of a planning cell within the Ministry of Defence, headed by an additional secretary, to "deal with the wider aspects of development planning."[4] The new system was intended to facilitate middle- and long-range defense planning and to maintain constant liaison with the Planning Commission and other ministries. It was expected that the defense effort would derive maximum advantage from the development effort by being able to place its priorities for consideration along with the overall economic and industrial plans being drawn up by the Planning Commission.

The experience with the 1964–1969 First Five Year Defence Plan displayed some of the basic problems of defense planning in India.

[4] The 1965–1966 annual report of the Defence Ministry noted that the "seriousness of the threat to the country's security posed by the unprovoked aggression in September 1965 and the ultimatum given by China in September 1965, emphasized the inevitable need to harness all resources of the country to the country's defence and for the defence effort to derive full sustenance from the country's economic development plans" (*Report: 1965–66*, [New Delhi: Ministry of Defence, Government of India], p. 5).

It became apparent, especially after the 1965 Indo-Pakistani war, that there were sharp differences between ordnance requirements during longer periods of peace and in shorter, emergency situations. There was a sudden increase in demand for arms and ammunition when war broke out, followed by a further demand for replenishing stocks in the immediate postwar period. Once stockpiling was completed, demand declined rapidly and the ordnance factories were faced with idle capacities.

One proposed solution—to maintain high production levels during times of peace and to stockpile armaments for all projected wartime needs—was considered economically undesirable. Such a policy carried the risk of obsolescence and implied tying up large amounts of capital that could be used elsewhere. Another proposal was to channel idle capacities in the defense production units into civilian production. This approach was adopted in part at the conclusion of the first defense plan when the Defence Ministry, under Swaran Singh, proposed the manufacture of civilian goods to maintain "both machine and human skill . . . in proper trim in order to be really effective in an emergency." The ministry felt that such a policy would "also enable the machines to be wasted out in a reasonable time, thereby ensuring their replacements by more modern and progressively more indigenous machinery."[5]

It also became clear that most major defense projects took more than five years to complete. The 1970–1971 report of the Defence Ministry observed that since the first defense plan had been attempted within the framework of a five-year period, it had been difficult to incorporate resource requirements for projects that took longer than five years into the plan. According to the ministry, "the cycle of development of weapons and equipment is generally five to seven years while that of more sophisticated weapons systems, advance aircraft and modern fighting ships, is even longer, say seven to ten years."[6]

In order to mitigate these drawbacks, future defense plans were made on a five-year "roll-on" basis. After a year was completed, an additional year was tagged on at the other end so that the services would always have before them a revised and updated five-year plan. This system was expected to facilitate the planning of projects that took five years or more as well as the revision, addition, or elim-

[5] *Report: 1969–70*, Ministry of Defence, p. 36.

[6] *Report. 1970–71*, Ministry of Defence, pp 6–7.

ination of projects, depending on the foreign exchange situation and the availability of domestic resources.

In the mid-1970s, a broader approach to security planning was manifested when the defense decision-making bodies were drawn into the overall political and economic decision-making bodies. The Defence Committee of the Cabinet (DCC) was transformed into the Political Affairs Committee of the Cabinet (PACC) and the Defence Minister's Committee (DMC) was replaced by the Defence Planning Committee (DPC).[7] This was in keeping with the new political perspective regarding security that sought to achieve a complexity of interrelated goals: defense against external enemies, the maintenance of internal peace and security, and the promotion of economic and social development.

The move has had mixed results for the economy and for the military. On the one hand, it has provided defense input in the national planning process and ensured that defense allocations do not undermine the economic and political stability of the nation. On the other hand, the new system has eroded military input into defense planning because representation on the PACC is through the defense minister, and the representation of the three service chiefs in the DPC tends to be overwhelmed by the presence of seven civilian departmental secretaries. In the present setup, the needs of defense tend to become submerged in larger national political, economic, and social issues. Although this may seem unfortunate from the military standpoint, it has been deemed essential for economic planning and stability.

The Location of Defense Production: Private versus Public Sectors

An economic issue of increasing importance in recent years has been the location of defense industries in the public and private sectors. The relevance of this stems from a shift in the government's economic policy from an earlier emphasis on a state-dominated economy to the current trend of giving the private sector a role in production. The question arises whether this shift should also be applied to defense production.

The location of defense industries assumes considerable importance because it involves the efficiency and effectiveness of weapons produced in India as compared to those deployed by India's adversaries. From an economic standpoint, it should be determined

[7] See section on defense decision making in chapter three for details.

which sector is best able to optimize weapons production. However, the issue of where defense industries should be located involves strong ideological and political arguments.

Following the proclamation of the Indian Republic in January 1950, control over defense and defense production was assigned to the central government through the Seventh Schedule of the Indian constitution. This clause provided that legislative jurisdiction over defense was to rest with Parliament, implying that political responsibility belonged to the government in power while Parliament as a whole was to play a guardian role. Although it is normal for modern states to retain the area of defense as a governmental responsibility, especially in the recruitment, organization, and conduct of the armed services, defense production is not always maintained exclusively in the public sector. In the United States and Western Europe the policy usually has been to fulfill defense needs for major weapons through private industry, granting contracts to the most economical and technically efficient bidder. The arms industry in the West is therefore part of the so-called mixed economy, in which private enterprise plays an important and economically profitable role. In India, the area of defense production was assigned to the state as a matter of public policy, although the Seventh Schedule only called for legislative jurisdiction and not state ownership of defense production.[8]

The decision to nationalize all defense industries was in keeping with the Congress government's often-repeated policy of establishing a socialist pattern of society. Indeed, the ideological decision to locate defense industries in the public sector was taken even before 1947. In 1938, independence in sight, the Congress had formed a National Planning Committee to draw up an economic program. The committee proposed a socialist path for India but with an important contributing role to be played by the private sector. Although there was considerable disagreement within the committee on adopting the socialist pattern, there was no dissent on the location of defense industries in the public sector.

The general socialist policy of the Congress party was further consolidated in two Industrial Policy Resolutions in 1948 and 1956 and by the creation of the Planning Commission in 1950.[9] The Planning Commission was to be the instrument whereby socialism would

[8] *Constitution of India*, List I, Union Government, Seventh Schedule.

[9] For a discussion of the aims and politics surrounding the creation of the Planning Commission, see R. J. Venkateswaran, *Cabinet Government in India* (London: George Allen & Unwin, 1967), pp. 109–121.

be implemented, but the resolutions governed the framework within which industrialization was to proceed, including the distribution of various types of industries between the public and private sectors.

The 1956 Industrial Policy Resolution, in particular, clearly outlined a fundamentally socialist policy. It envisaged progressive state control and ownership of almost all major industries in the long run.[10] Again, although there was controversy over the economic efficiency of state-controlled and -owned industries, deliberation on the location of defense industries was excluded from public debate and scrutiny; from the outset, all defense industries fell to the public sector. In addition, certain areas of production were classified as "strategic" industries because they were considered vital to the defense effort in times of military danger. These too could be taken over by the government on short notice by emergency decree.

Disagreement about the role of the Indian public and private sectors in industrial production, including defense production, has traditionally centered on two related aspects. The first has involved the right of private individuals who constitute the very small business and industrial community in India to own capital and make profits in a country where most of the population is very poor. The second and more relevant issue has focused on the productivity and efficiency of public sector undertakings as opposed to industries in the private sector.

In the first case, it has been clear that industries directly producing weapons and military equipment for the armed services would remain in the public sector. There has been some ambiguity, however, in the area of noncombat items such as clothing, provisions, general stores, and component and spare parts that become part of the final weapon product. There also have been other grey areas, such as industries related to the defense effort in the areas of communications and electronic equipment, railway rolling stock, transportation vehicles, and materials that could be diverted to defense needs.

The manufacture and supply of clothing and general stores have almost exclusively been assigned to the director-general of ordnance factories in the Department of Defence Production. The economic value of this policy is somewhat dubious; the private sector is

[10] For a right-wing political discussion of this policy, see B. R. Shenoy, *Indian Economic Policy* (Bombay: Popular Prakashan, 1968), pp. 49–56. See also D. S. Kushavala, "Role of Industrial Policy in Economic Development," in *Indian Economic*

better suited to produce such goods because it already caters to the general population. Private business interests control virtually all of the textile and clothing industries. However, one interpretation of the 1956 Industrial Policy Resolution appeared to require that the manufacture of such goods for the military should fall to the governmental sector despite the lesser economic advantage held by the state. Consequently, several clothing factories and the Cable Factory at Chandigarh, all under the Ministry of Defence, produce a large variety of winter and summer clothing and items such as field cables, tents and tent poles, and various types of parachutes and floats.[11]

On the other hand, a substantial quantity of the innumerable component and spare parts that go into the major items of equipment manufactured at ordnance factories and the public sector undertakings managed by the Defence Ministry had to be obtained from other civilian manufacturing concerns in the public sector as well as from the private sector. A strict interpretation of the 1956 resolution would require that these items also be produced in the public sector. But the uneconomical nature of defense manufacture of this diversity of goods was evident, and some collaboration with the private sector was therefore inevitable.

Until 1965, no clear system and policy existed in the Ministry of Defence for tapping the civilian market to supply such parts. In that year a Department of Defence Supplies was created in the Ministry of Defence primarily to promote the indigenous production of previously imported defense stores as well as new stores required by the government-run defense factories.[12] Additionally, several technical committees were set up in the new department to locate indigenous sources of items required by defense public sector undertakings. These committees covered aeronautical, armament, electronic and electrical engineering, marine, medical, vehicle, and general stores needs, and components for the manufacture of the Vijayanta tank. The stores included such items as mounting tripods, sight periscopes, gear boxes, generating sets, switchboards, antenna equipment, gaskets of all kinds, water tanks, Bailey Bridge steel components, launching pendants, helmets, boots, and an assortment of items for vehicles and tanks including tires, carburetors, pistons, and radiators.

Thought and Development, ed. Ashok V. Bhuleshkar (London: C. Hurst, 1969), pp. 214–27.

[11] *Report: 1972–73*, Ministry of Defence, pp. 32–33.

[12] *Report: 1972–73*, Ministry of Defence, p. 92.

In determining raw materials and supplies that were of "strategic" importance in the sense that they directly or indirectly backed the defense effort, the concept was broad enough to encompass all major industries as well as agriculture. The Seventh Schedule had provided that "industries declared by Parliament by law to be necessary for the purpose of defence or for the prosecution of war" could be taken over by the government if needed. Since most major products such as iron, steel, petroleum products, and heavy machine tools already fell under the public sector (but nondefense) corporations, the role of the private sector (and consequently the government's need to intervene) was quite limited.

The government's refusal to allow the private sector to participate in defense production in any meaningful way was explained rather inadquately as far back as 1965, when the Defence Ministry said that the manufacture of armaments called for "special skills and capacities to work to close tolerances." The ministry's statement concluded that "while it is generally recognized that the capacity for filling of ammunition or for the manufacture of special propellants and explosives has to be in the Defence sector, it is not often appreciated that the capacity for forging and the high standards of machining required for ammunition shells has also be to in the Defence sector as the civil industry does not ordinarily need or have such capacity."[13] This statement would suggest that one reason for rejecting private industrial participation in defense production was its inability to meet complex and rigid engineering specifications demanded by the Ministry of Defence. The statement would also imply that in subcontracting the manufacture of defense items to the private sector, the Defence Ministry would not have high standards and strict time schedules. Neither implication seems borne out, however.

In the first instance, India's leading industrial firms, such as Tatas, Birlas, Mafatlals, Dalmias, and Walchands, have the entrepreneurial talent, technology, and efficiency to carry out the manufacture of most of the weapons now produced in the public sector. Their capacity to deliver the goods in accordance with the rigid specifications demanded by the Defence Ministry cannot be seriously doubted. Even capacities that do not presently exist in the private sector would no doubt be created swiftly if sufficient and continuous defense orders were forthcoming. For example, both Tatas and Birlas produce cars, trucks, and other vehicular goods. The Tata truck is widely accepted both in India as well as in Middle East-

[13] *Report: 1965–66*, Ministry of Defence, pp. 32–33.

ern and Southeast Asian markets as being of a very high quality, rivaling that of its original collaborator, Mercedes-Benz. Walchand Industries was one of the pioneers in India in the manufacture of aviation equipment; it was the founder of Hindustan Aircraft Ltd. in Bangladore in the early 1940s and managed the company for a decade before the government nationalized it and put it under the Defence Ministry. There can be little doubt about the private sector's relative economic viability and technological and engineering capabilities.

In the second instance, governmental control over defense items subcontracted to private industries is likely to be nearly total due to competition from rival manufacturers anxious to secure the same contract. The willingness to make concessions by competing firms should enable the Defence Ministry to demand greater control in the production process. Indeed, the engineering division of the Defence Ministry usually maintains close supervision throughout the manufacturing process, assigning inspectors in many cases to make weekly visits or to work continuously with the private manufacturer who is fulfilling the defense contract. Although high rejection rates of items may be required due to failure to meet rigid specifications, this is probably also the case of manufacturing establishments in the public sector, whose efficiency is usually lower than that of most private sector industries.

The second debate arising from the consignment of all defense industries to the public sector concerns the efficiency of public sector corporations versus industries in the private sector. Because the decision to distribute different industries to different sectors was essentially an ideological one, its economic rationale and consequences were relegated to a position of secondary importance. Yet the question remained whether the quality of weapons needed by the armed services could be sacrificed for ideological or social reasons. From the military perspective, the practice of becoming self-sufficient in civilian goods, even if these were below the standard of those available in the international market, could not be applied to the defense sector. Although it might be tolerable for an Indian-made locomotive or transport vehicle to break down during civilian use, a gun that did not fire, a tank that had engine trouble, or combat aircraft that had to be grounded constantly or that crashed even before they challenged the enemy could destroy the effectiveness and morale of the military during times of war. Defense equipment had to match the quality and performance of similar weapons acquired by India's adversaries.

The primary fears of defense undertakings in the public sector are those that concern all such governmental enterprises: the alleged economic inefficiency of state-run corporations facing little or no competition and the bureaucratic politics within them that make the prospects of improved productivity even less likely. There is also the difficulty of establishing or even defining "social returns" on the capital invested if justification cannot be found on strictly economic grounds. These general problems become more difficult in the case of defense, where economic criteria can be easily rejected for the sake of national security.

The number of public sector undertakings in India, including those under the Ministry of Defence, had risen from 5 at the start of the First Five Year Plan in 1950 to 176 in 1979 (excluding insurance corporations). The investment in these enterprises during this period rose from Rs. 29 crores to Rs. 15,602 crores.[14] The distribution of investment between the public and private sector had increased in ratio from about 45:55 in the First Five Year Plan to about 65:35 at the end of the Fourth Plan. By 1980, there were nine public sector corporations under the Ministry of Defence: Hindustan Aeronautics Ltd., Bharat Electronics, Mazagon Docks, Garden Reach Shipbuilders and Engineers, Goa Shipyard, Bharat Earthmovers, Bharat Dynamics, Praga Tools Ltd., and Mishra Dhatu Nigam. The total investment in these public sector undertakings was Rs. 632 crores in 1982, of which Rs. 244 crores was the paid-up share capital and Rs. 388 crores was in the form of loans. (See table 6.1.) This constitutes about 20 percent of total investment in public sector undertakings.[15]

On the whole, a comparison of the rates of return between public sector and private sector corporations indicates that in both absolute and relative terms, the former has tended to be less profitable. An analysis of balance-sheet data by the Reserve Bank of India shows that central government companies have been averaging a profit rate (usually under monopolistic conditions) of about 4.4 percent and state government companies of about 2.2 percent between 1970 and 1975.[16] In comparison, the average profit rate for the private corporate sector has been about 11 percent. Among the reasons given for the public sector's low profitability have been government

[14] *India: 1981* (New Delhi: Publications Division, Ministry of Information and Broadcasting, 1981), p. 229.

[15] *India: 1981*, p. 39. See also *Report: 1979–80*, Ministry of Defence, pp. 34–36.

[16] Raj Krishna, "The Economic Development of India," *Scientific American* 243, no. 3 (September 1980):174.

TABLE 6.1

Investment, Production, and Sales in Defense Public Sector, 1980–1983 (in Rs. crores)

	1980–1981		*1981–1982*		*1982–1983 (Estimated)*	
Enterprise	*Paid-up Share Capital*	*Govt Loans*	*Paid-up Share Capital*	*Govt. Loans*	*Paid-up Share Capital*	*Govt. Loans*
HAL	60.00	105.00	60.00	142.67	60.00	150.00
BEL	13.00	15.60	13.50	16.23	16.50	20.63
BEML	12.00	19.37	12.00	19.92	19.00	22.55
MDL	13.50	16.26	26.00	32.57	51.00	56.20
GRSE	24.90	37.57	27.40	44.57	29.90	54.07
GSL	1.92	1.56	1.92	1.91	1.92	3.17
PTL	6.99	5.46	6.99	4.40	7.24	4.24
BDL	2.60	2.59	3.40	3.68	5.42	2.62
MDNL	45.00	59.35	45.00	69.98	52.60	74.68
Totals	179.91	262.85	196.26	335.83	243.58	388.11

	1980–1981		*1981–1982*		*1982–1983 (Estimated)*	
Enterprise	*Value of Production*	*Sales*	*Value of Production*	*Sales*	*Value of Production*	*Sales*
HAL	158.41	166.25	275.03	226.64	333.80	265.86
BEL	80.83	68.91	127.17	128.44	148.93	142.00
BEML	98.20	87.91	225.20	221.93	317.75	295.60
MDL	86.95	89.55	126.77	111.54	145.79	110.90
GRSE	39.02	38.14	50.33	42.38	49.40	64.26
GSL	7.77	6.02	8.52	1.51	13.50	18.06
PTL	8.50	9.32	8.30	7.58	11.87	11.29
BDL	4.04	5.04	3.15	3.19	1.62	3.73
MDNL	1.60	1.60	4.91	3.20	10.57	8.30
Totals	485.32	472.74	829.38	746.41	1028.23	920.00

SOURCE *Report· 1982–83* (New Delhi. Ministry of Defence, Government of India, 1983), pp 48–50.

NOTE Data given as reported in source. Abbreviations: HAL–Hindustan Aeronautics Ltd.; BEL–Bharat Electronics Ltd.; BEML–Bharat Earthmovers Ltd.; MDL–Mazagon Docks Ltd.; GSL–Goa Shipyards Ltd.; PTL–Praga Tools Ltd.; GRSE–Garden Reach Shipbuilders and Engineers; BDL–Bharat Dynamics Ltd., MDNL–Mishra Dhatu Nigam Ltd.

pricing policies; capital structures that tended to be half in the form of loans—resulting in heavy interest burdens; lack of financial and economic objectives; excessive expenditures on staff amenities; poor labor-management relations; and over-provisioning for depreciation of plant and machinery.

Most of these problems apply to the defense public sector undertakings as well. However, the economic performance of defense undertakings has proved to be above the general average of other governmental enterprises. Profits in the defense sector have ranged from about 2 percent for Hindustan Aeronautics Ltd. in 1980 to more than 12 percent for Bharat Electronics. (See table 6.2.) But note that HAL is the largest defense sector corporation; its paid-up capital exceeds the combined paid-up capital of the other undertakings. The earlier low rate of profitability is now less significant since HAL dividends rose to 10 percent in 1981–1982.

Just as considerations of equity and the common social good called for the location of all major industries in the public sector regardless of economic criteria such as optimizing cost, quality, and production time (if we assume that greater private sector participation could produce better results), so too ethical considerations seemed to suggest the need to avoid making private profit in national defense enterprises. However, in view of the present ideological shift in India toward "liberalization" and "pragmatism" (euphemisms for industrial decontrol and greater private sector production), should this also be extended to the defense sector?

Most of the arguments advanced for the shift in policy toward increased private sector participation in civilian production may also be advanced in the case of defense production. Despite bureaucratic, economic, and technological restrictions, the hitherto semi-handcuffed private sector in India has nevertheless been able to compete effectively in overseas markets in an array of manufactured items, including highly sophisticated engineering goods. The lowering of restrictions under the new policy is expected to accelerate the international competitiveness of private Indian firms. The Indian private sector has already made considerable progress in the high technology industries and possesses a sophisticated capability in the automotive industry as well. Private sector capability thus carries a significant defense potential, which will increase as capability expands under the policy of liberalization. Under these circumstances, if substantial private sector participation in weapons production is permitted, it could lead to the optimization of time, cost, and quality—the essential demands of defense production.

TABLE 6.2
Performance and Profitability of Defense Public Sector, 1980–1983

a. Working Results (in Rs. crores)	*1980–81*	*1981–82*	*Estimated 1982–83*
Value of production	485.32	829.38	1028.23
Turnover	472.74	746.41	920.00
Profit before tax	20.08	67.25	61.74
Return on turnover	4.25%	9.01%	6.71%

b. Profitability (in Rs crores) *Enterprise*	*1980–81*	*1981–82*	*Estimated 1982–83*
HAL	4.22	29.64	31.19
BEL	8.93	20.13	20.99
BEML	9.87	29.88	39.89
MDL	4.72	5.63	7.50
GSL	0.49	0.36	0.50
PTL	1.26	0.92	1.12
Total (a)	29.49	86.56	101.19
GRSE	−7.98	−12.53	−28.85
BDL	−1.43	−1.48	−3.11
MDNL		−5.30	−7.49
Total (b)	−19.41	−19.31	−39.45
Total (a) + (b)	20.08	67.25	61.74

c. Dividends Paid (in Rs. lakhs and % of Unit Share) *Enterprise*	*1980–81*	*1981–82*
HAL	120.00 (2%)	600.00 (10%)
BEL	142.00 (12%)	158.00 (12%)
BEML	72.00 (6%)	72.00 (6%)
MDL	94.33 (11%)	128.59 (10%)
GSL	15.39 (8%)	11.55 (6%)
Totals	443.72	970.14

SOURCE: *Report: 1982–83* (New Delhi: Ministry of Defence, Government of India), p. 49–52.
NOTE For explanation of enterprise names, see table 6.1.

Two major criticisms may be advanced against locating some or most of defense production in the private sector. First, there might be a tendency on the part of private firms to escalate the cost of weapons programs beyond reasonable profit levels because they are aware that the government will pay the bill. A counterargument could be made that contract bids are likely to be competitive and therefore the cheapest estimate possible will be submitted, but this may not happen in India because it has few dominating private industrial houses. Costs may also escalate after a contract has been concluded. However, this is likely to be resolved in the long run or the contracting firm will face the prospect of losing future governmental defense contracts.

The second argument concerns the undue influence that may be gained by the military-industrial complex in the making of foreign policy. Private sector defense production may tempt industrialists to support political candidates who favor aggressive foreign policies that would increase the demand for weapons. Eventually, this could produce a set of national economic priorities that excessively favored war industries over investments in consumer-oriented goods and services and diverted resources and human endeavor away from more urgent social objectives. There may be some truth to this argument, although it is not easily demonstrated. However, given the poverty of the electoral masses and the comparatively insignificant size of the private business community in India, the composition of Parliament would remain relatively the same and enable it to exercise control over the direction of India's foreign and defense policies.

II. Defense and the Economy

Defense Spending, Investable Savings, and Social Expenditure

Indian defense spending as a percentage of the GNP has varied between 3 and 4 percent since the outbreak of the Sino-Indian war in 1962. Highs of 3.8 to 4.0 percent were reached in 1963, 1966, 1971, 1972, and 1979.[17] Except for 1980, when a major modernization drive for the armed forces was under way, the other peaks were reached during or immediately after the wars with China and Pakistan. As a percentage of central government expenditure, defense spending has declined from an average of about 31 percent in 1964 to about 20 percent in 1970s and early 1980s.[18] (See table 5.5.) The present central budget allocations therefore compare favorably with the level of about 17 percent that had generally prevailed in the 1950s.

These figures would not suggest an excessive level of defense

[17] Figures derived from the *Institute for Defence Studies and Analyses Journal* (New Delhi) 4, no. 4 (April 1972):434; *The Military Balance, 1981–82* (London: International Institute for Strategic Studies, 1981); and *India. 1981*.

[18] *India: 1981*, p. 34.

spending in India. The average defense spending of approximately 3.5 percent of GNP between 1962 and 1982 appears reasonable when compared with 10 percent in China and Taiwan, 7 percent in Pakistan, 5 percent in South Korea, 15 percent in the Soviet Union, and 6 percent in the United States.[19] It seems even more insignificant when compared with countries like Israel, Egypt, Saudi Arabia, Iran, and Iraq, whose defense spending has been 20–35 percent of the GNP at various times. There are differences, of course, between India and Third World countries that have much higher relative levels of defense spending: the oil-rich countries have substantial petrodollar reserves that may be readily expended on arms imports, and nations such as Israel, Egypt, Taiwan, and Pakistan have been recipients of a much higher per capita inflow of economic and military aid. Thus the level of defense spending relative to the GNP may not be an accurate reflection of the burden of defense.

One argument advanced in India in the mid-1960s against increased defense spending was based on the level of investable savings available in developing countries. Unlike the western industrialized nations and Japan, which have highly developed and extensive financial institutions capable of tapping substantial savings for investment, poor countries are characterized by high consumption pattern economies and a general lack of banking and financial institutions. The essential point in the early 1960s was that although Indian defense spending was less than 4 percent of the GNP, this constituted a substantial part of the savings (about 15 percent of the GNP) that could be generated in India at the time.[20] The diversion of more than one-fourth of the national savings to the defense sector could considerably affect the Indian GNP growth rate. Ultimately a lower growth rate, or even a negative growth rate, would mean that the absolute amount available for defense at the same rate of about 4 percent of the GNP would be less than what might have been available with a lower rate of defense spending but a higher growth rate.

Whether or not the substantial usurpation by the defense sector of the available investable savings in India is the cause of lower GNP growth rates will be discussed later. In the present context, it is interesting to note that although defense spending has remained nearly constant at about 3 to 4 percent of the GNP, the level of in-

[19] Figures from various issues of *The Military Balance*, 1971–72 to 1982–83.

[20] See Raju G. C. Thomas, *The Defence of India: A Budgetary Perspective of Strategy and Politics* (New Delhi: Macmillan Co. of India, 1978; Columbia, Mo.: South Asia Books, 1978), p. 134.

vestable savings has risen steadily from about 10 percent of the GNP in the early 1950s to 17 percent in the mid- to late 1960s, 20 percent in the 1970s, to a high of 24 percent in 1981.[21] The relative share of the defense sector of the total investable savings generated has, in effect, declined over the last two decades. Indeed, despite the increasing absolute levels in defense spending, the annual defense budget now absorbs the same amount of the central government budget as it did in the 1950s—down from over 30 percent in the aftermath of the 1962 Sino-Indian war to about 20 percent in the early 1980s.

However, the level of investable savings generated in India does not compare favorably with that of China. Both India and China are reported to have started with a savings level of about 15–17 percent of the GNP in the early 1960s.[22] Although India now generates 24 percent, China generates 31 percent, a level that compares favorably with the advanced industrialized nations. China's performance may appear all the more creditable given the fact that only 1 percent of China's investment was derived from external capital flows compared to about 3 percent for India. On the other hand, India can claim that its percentage of investable savings diverted to defense is proportionately lower than that diverted by China: India's defense allocation of 3–4 percent of the GNP out of a savings of 24 percent of the GNP (one-sixth) compares to an allocation of about 10 percent out of a savings of 31 percent (about one-third) in the case of China. Although these figures suggest the differing economic and security concerns on either side of the Himalayas, ultimately the question is to what extent defense diversions have adversely affected development and economic growth in the two countries.

Overall, a savings of 24 percent of the GNP is quite remarkable for a developing nation in which more than two-thirds of the population (500 million people out of a total of 750 million in 1985) still live at or below subsistence level. India's per capita GNP was about $250 in 1984; it ranked as one of the poorest countries in the world. According to one perspective, this savings feat has become possible only because of the national commitment to defense. Reference is made to the national call for economic sacrifice in the aftermath of the 1962 war that enabled the Ministry of Finance to siphon off ex-

[21] Figures obtained from Krishna, "The Economic Development of India," p. 172. See also the World Bank's *World Development Report, 1981* (New York: Oxford University Press, 1982), p. 142.

[22] See Dilip Mukerjee, "Growth Rates in China and India," *Times of India*, October 16, 1982.

tra savings for the defense effort, momentum that has since been maintained. Whatever the validity of this argument, the relative diversion of available resources to defense has continued to decrease as the government continued to tap higher levels of savings.

Defense spending is generally perceived as a form of "service" expenditure, in which the main benefit is the intangible commodity known as "security" (in India, defense expenditure is classified as "non-plan" expenditure). It is therefore useful to compare this service with two other major services: health and education. The issue is whether the service of defense is being given inordinate attention or is being obtained at the cost of these other two vital services, and it is here that India appears less favorable compared with most other major developing nations. Although, for example, 1982 military expenditure in India was lower, at $8 per capita, than most other states in the world, China nevetheless spent $4 per capita on health in comparison to India's $3 per capita, and $9 per capita on education compared to India's $8.[23] India's per capita expenditure appears even more questionable when compared with that of Brazil ($37 and $93), Mexico ($13 and $113), or Taiwan ($67 and $96) and there can be little satisfaction to know that in Pakistan these figures appear to be equal or worse. However, such figures merely reflect the much lower per capita income in India. On a per capita basis, India spends as much on education as on the military, and a little less than half as much on health. In contrast, the per capita allocations of China, Taiwan, and Pakistan to education and health appear much less favorable in terms of their per capita allocations to defense.

Two related questions remain. How has the substantial increase in savings from one decade to another affected the Indian GNP growth rate? Would an increase in the allocation to defense (say from the present 3.5 percent of the GNP to about 5 or 6 percent) adversely affect the Indian economic growth rate?

Although savings increased measurably from about 15 percent of the GNP in the 1960s to about 20 percent in the early 1970s and defense spending remained around 3.5 percent of the GNP through both decades, the GNP also maintained an average of 3.5 percent rate of growth during this period. However, this rate increased to an average of 5 percent in the Fifth and Sixth Five Year Plans from 1975 to 1984, and the Planning Commission now estimates that an

[23] Figures drawn from Ruth Leger Sivard, *World Military and Social Expenditures, 1985* (Leesburg, Va.: World Priorities, 1985).

average 6 percent growth rate will be maintained during the 1985–1990 Seventh Five Year Plan.[24]

In the past, the reasons for the discrepancy could be found in economic inefficiencies and waste, the vagaries of the monsoon rains that often wreak havoc on food production in certain years, and international factors such as the sudden spurt in oil prices. Because of such bureaucratic failures, natural calamities, and unexpected events, India's annual growth rates have fluctuated considerably. In the 1960s, the GNP growth rate was 2.6 percent in 1962–1963, 7.3 percent in 1964–1965, −5.2 percent in 1965–1966, 9.9 percent in 1967–1968, and 2.5 percent in 1968–1969.[25] Similarly, in the 1970s, the GNP growth rate swung from −1.1 percent in 1972–1973 to 9.6 percent in 1975–1976 and −4.8 percent in 1979–1980. In the early 1980s growth rates showed signs of remaining high and commensurate with the high savings rate of 24 percent: 7.5 percent growth in 1980–1981, and 5.2 percent in 1981–1982. However, the growth rate fell to 1.8 percent in 1982–1983 because of unfavorable monsoon rains.

However, the trend of annual GNP growth rates over the last twenty years does not indicate an empirical linkage with the steady growth of the savings rate. The present healthy performance of the Indian economy may be suddenly reversed if there are serious economic dislocations arising from war, monsoon failures, foreign exchange crises, industrial unrest, bottlenecks in the flow of goods and services, or inefficiency and mismanagement of the economy.

The degree to which defense allocations contribute to the ups and downs of Indian GNP growth rates is unclear and the benefits or losses related to defense spending remain controversial, as discussed below.

Perspectives in the Defense-Development Debate

The relationship between defense and development suggests a dual paradox. The first pertains to the belief that more defense spending may not buy more security but may in fact lead to less security if rival states proceed to match or exceed India's increased defense alloca-

[24] See article by a member of the Indian Planning Commission, A. M. Khusro, "Some Innovations in the Seventh Five Year Plan," *Indian and Foreign Review*, August 31, 1984, pp. 5–7.

[25] Figures derived from K. Subrahmanyam, *Defence and Development* (Calcutta: Minerva Associates, 1973), pp. 2–3; *Christian Science Monitor*, special supplement on India, July 30, 1982; and *India: 1981*, p. 34; and *The Times of India Directory and Yearbook, 1984* (Bombay: Times of India Press, n.d.), p. 288.

tions. India has generally claimed that Pakistan was responsible for initiating the arms race on the subcontinent when it obtained large amounts of sophisticated arms under SEATO and CENTO in the 1950s and again more recently after the Soviet invasion of Afghanistan when it sought and obtained a $2.5 billion American arms sales package that included the advanced F-16 fighters.[26] However, Pakistan perceives India to have triggered the arms race with its decisions to procure the Anglo-French Jaguar aircraft in the late 1970s and its $1.6 billion arms package with the Soviet Union in 1980. Thus, successful Indian arms deals were responsible in part for the Pakistani acquisition of even more sophisticated American weapons. In turn, this has produced the Indian decision to purchase the Mirage-2000 from France at the unprecedented cost of $3.3 billion. Since the Mirage-2000s are not likely to be delivered until the late 1980s, the accelerating cycle of the arms race on the subcontinent has meant considerable cost for India but little appreciable increase in its sense of security.

A similar concern is the Indian nuclear option. Those opposed to acquiring nuclear weapons argue that possession of such weapons would not only escalate the Chinese nuclear threat but would also trigger a Pakistani attempt to acquire such weapons. India would then be involved in a costly three-way nuclear arms race resulting in less rather than more security.

The second paradox, which is of greater relevance in the economic debate, rests on the contention that less defense spending may not necessarily produce more economic development. On the contrary, defense programs may carry certain beneficial aspects for development. This perspective is opposed to the traditional and more widespread belief that maintains that military expenditures usurp resources from the development program and thereby curtail economic growth.

With some variations, the debate in India corresponds to these fundamentally opposing views. The traditional perspective that emphasizes the opportunity costs of defense programs has generally prevailed since Independence and remains an important contending viewpoint. Only during the last decade has the opposing belief that emphasizes the complementary nature of defense and devel-

[26] For Indian reactions to the U.S.-Pakistan arms agreement, see *Times of India*, March 31 and May 18, 1982. For reactions to the Indo-Soviet arms deal, see *Statesman*, May 29, 1980; *New York Times*, May 30, 1980; and *Overseas Hindustan Times*, June 12, 1980.

opment begun to gain recognition. Four variations of these two opposing beliefs may be found in India.

First Perspective. Before the 1962 Sino-Indian war, the belief existed in India that not only did defense usurp development resources but that it would be morally wrong for a poor nation such as India to be engaged in a costly arms buildup. Gandhian notions of moral force and peaceful resistance were implicit in the early Indian policy of nonalignment. Nehru believed that nonalignment would resolve the dual objectives of obtaining security and economic development.[27] By not getting involved in the cold war between East and West and by offering a policy of friendship to both sides, India hoped to increase its sense of security without heavy defense spending and to enable the country to utilize a maximum of its resources for development. The paramount claims of development over defense in the 1950s is reflected in the annual allocation of less than 2 percent of the GNP to defense (with marginal exceptions of 1957 and 1958 when it rose to 2.4 percent and 2.2 percent).

Second Perspective. The second belief, which gained ascendancy among government decision makers after the 1962 Sino-Indian war, generally acknowledged the first belief that defense and development were competitive. But new military circumstances seemed to call for priorities. Development only made sense if it could be protected from outside aggression, even if this implied less development because of the greater diversion of resources to defense.[28] The changes in policy and outlook after 1962 were more political than economic. Although there was a more conscious and careful

[27] The assumption that nonalignment provided the best prospect for peace and was therefore a prerequisite for economic development was elaborated by Prime Minister Jawaharlal Nehru during a visit to Sri Lanka in 1954. Nehru stated that it would be a "tragedy of infinite magnitude if we should be checked and baulked and our (economic) policy should be set at naught because of the troubles and quarrels of others" (*Hindu,* May 2, 1954). See also Raju G. C. Thomas, "Nonalignment and Indian Security: Nehru's Rationale and Legacy," *Journal of Strategic Studies* 2, no. 2 (September 1979): 153–71.

[28] This attitude was best expressed by then Foreign Minister Morarji Desai, when he declared that national sacrifice could never be enough when it came to defending the nation's territorial integrity. According to Desai, "defence is our first charge and whatever our conditions, we should spend all for our defence. If we have to spend every paise to defend our freedom and to fight aggression, we will have to do" (*Hindu,* June 15, 1967). For similar pronouncements by Prime Minister Lal Bahadur Shastri and Planning Minister Ashok Mehta, see *Indian Express,* October 11 and 19, 1965; and the *Statesman,* May 15, 1966.

evaluation of the consequences of increased defense expenditures on development, the claims of defense were now given more importance than those of development except in critical areas where development plans were considered vital to the defense effort.

This changed attitude—which was not unexpected in the suddenly changed military conditions of the 1960s—acknowledged the adverse effects of defense spending on the economy as well as the long-term dependence of defense on development. The resources that could be diverted to defense were dependent, after all, on the nation's industrial and technological progress. The greater the GNP, the greater the defense spending that became possible. Consequently, diversions to defense had to be undertaken without seriously jeopardizing the long-term development programs since this would reduce the absolute size of the defense program itself in the future. In other words, a higher level of defense spending at (say) 7 percent of the GNP that produced a lower GNP growth rate of 2 percent would imply a lower absolute amount of resources available for defense at the same rate of defense spending than (say) 2 percent defense spending that produced an economic growth rate of 7 percent. In the long run, higher levels of defense spending become self-defeating for the defense program itself. As is apparent in the trend of defense allocations over the two decades since the 1962 war with China, India's defense policy makers considered it wise to maintain the level of defense spending at about 3.5 percent of the GNP.

Third Perspective. From the early 1970s on, the Indian approach to defense planning reflects a subtle but significant change in the perspective of the relationship between defense and development. Although the level of defense spending remained at about 3.5 percent of the GNP during this decade, efforts were now made both to minimize the adverse effects of defense spending and to implement it in such a way to produce beneficial effects for the development program. The coordination of defense and development programs, based on the belief that they complemented rather than competed with each other, was the long-term outcome of defense planning in India that commenced after the 1962 Sino-Indian war. This approach initially stemmed from the awareness of and attempt to avoid the detrimental effects of larger allocations of funds to the defense sector. Twenty years of defense planning subsequently resulted in the "structural unity" of the defense and civilian sectors of the economy.[29]

Defense production units under the Ministry of Defence in the mid-1980s are not only involved with the production of military goods but also with a variety of civilian goods. Indeed, except for the ordnance factories that manufacture small arms, ammunition, military vehicles, military clothing, and heavier combat items such as tanks and armored personnel carriers, almost all the major public sector undertakings under the Ministry of Defence are heavily engaged in civilian production. For instance, Hindustan Aeronautics Ltd. manufactures the HS-748 transport planes that are used by the domestic Indian Airlines and the small Basant aircraft for agricultural use; Bharat Electronics produces a wide variety of civilian goods used in the communications and medical fields including commercial television; Bharat Earthmovers produces earthmoving tractors and equipment that are used in the construction and other civil engineering industries; and Mazagon Docks, Garden Reach Shipbuilders and Engineers, and Goa Shipyards produce oil rigs, barges, tugs, and cargo vessels and provide repair facilities for both military and commercial vessels.[30]

Conversely, there are civilian production units in both the public and private sectors that now contribute to a variety of military needs. The most significant example is the nuclear energy program, which though primarily intended for civilian purposes, could on short notice produce weapons for military purpose at an acceptable incremental cost. Other manufacturing units such as Hindustan Machine Tools and Instrumentations Ltd. in the public sector and units owned by the leading private business houses such as Tatas, Birlas, Mafatlals, and Dalmias have contributed in varying capacities to the defense sector despite the fact that the nature of their business is civilian. In fact, the Department of Defence Production within the Ministry of Defence is actively engaged in seeking out private civilian units that may contribute to defense production.[31] This complementarity of production between the military and civilian sectors is believed to provide economic benefits for both, especially in the automotive, shipbuilding, aeronautics, electronics, and

[29] See K. Subrahmanyam, "Planning Defence Production: Integral to Industrial Growth," *Times of India*, January 24, 1980.

[30] See *Report: 1979–80*, Ministry of Defence, pp. 27–43.

[31] For a description of how this department works, see chapter entitled "Department of Defence Supplies," in *Report: 1979–80*, Ministry of Defence, pp. 59–61.

construction industries. Any sudden withdrawal of defense spending and allocations is perceived to affect the efficiency of both civilian and military production.

The second and third perspectives on the relationship between defense and development have coexisted since the early 1970s. There appears to be no expressed change by a government official or politician that marks the beginning of a new attitude on these issues; the situation in the mid-1980s is more one of ambivalence between the second and third approaches. The perspective of the 1960s continues to be displayed in the cautious allocations to defense, while the perspective of the 1970s is seen in the integrated structure of the defense and civilian economies. Nevertheless, there is a fundamental difference in the underlying philosophies of the two. The third perspective projects a net positive effect of defense programs on development even if there is some competition for scarce resources. This is quite unlike the first and second positions, which predict adverse consequences of defense spending for development programs.

Fourth Perspective. Finally, a fourth approach may be noted that is basically an extension of the third. Whereas the third view suggested that there may be marginal and incidental benefits from defense programs for the civilian economy, the fourth considers this almost certain. Indeed, defense programs are considered to be essential for the development effort in a heterogeneous state like India. This emphasizes the increased "aggregate demand" provided by defense that taps idle resources in a developing country as well as promoting the civilian industrial spinoffs of defense production and also points to military contributions to the national psyche that encourage saving, investment, and the willingness to endure economic hardships for national security and the long-term gain. The Indian military symbolizes national unity and national discipline. It represents a cross section of the population, demonstrates the success of national integration, and sets an example for other sectors of the political system. Without this sense of national unity, development itself would be frustrated because of constant political agitation and fissiparous tendencies.

The fourth outlook does not follow chronologically and sequentially from the second and third perspectives but has existed among some sectors of the military, civilian defense specialists, and the national media since the 1962 Sino-Indian war. It is not a dominant view. With the exception of the psychological dimension emphasizing the beneficial effects for national unity, the economic components of this view are much less discernible or even indirectly reflected in official government policy. It has been expressed mainly by defense specialists, especially K. Subrahmanyam of the Institute for Defence Studies and Analyses in New Delhi, and by former military service chiefs in their individual and unofficial capacities, notably Gen. J. N. Chaudhuri.

For example General Chaudhuri, who had been the army chief of staff during the 1965 Indo-Pakistani war, wrote in 1973 that this "myth" of development programs being usurped by defense was still evident among Indian ministers, the civil service bureaucracies, and the public whenever they attributed India's slow economic growth to the nonproductive expenditure incurred in defense programs.[32] Chaudhuri instead pointed out the high rates of economic growth in countries like Israel, Taiwan, and South Korea that also had high rates of defense expenditure, concluding that "economic progress is influenced by many factors and to try to put the major portion of the blame for our slow progress on our defence expenditure is, to say the least, both unfair and uninformed."

By far the clearest statement of this line of reasoning came from Subrahmanyam, who declared that an empirical investigation into the defense expenditures of several nations during the last twenty years would reveal that higher military spending was invariably accompanied by high rates of economic growth. This included both Communist and non-Communist countries in Europe and Asia. Subrahmanyam said that "to derive a negative correlation between a low defence expenditure and high economic growth rates by taking Japan alone is not statistically meaningful. If Japan is to be cited as a single example of low defence expenditure and high economic growth, one could cite more than thirty other industrialized countries which have had a high defence expenditure and high economic growth rate."[33]

Arguments Underlying the Defense-Development Debate

Two basic issues are behind the principal arguments advanced against large-scale allocations to defense: (a) Defense allocations divert scarce resources, foreign exchange, and skilled manpower from the civilian development programs. Consequently, defense

[32] General J. N. Chaudjuri (Retd.), "The Defence Expenditure," *Indian Express*, May 11, 1973.

[33] *Indian Express*, May 19, 1973.

slows down the economic growth rate. (b) Defense programs produce adverse effects on the economy, causing structural maladjustments and price inflation.

Opportunity Costs. The first argument, which emphasizes the opportunity costs of defense, assumes an inverse relationship between defense and development. Only so much iron, coal, petroleum and other basic resources are available every year. Money spent on defense is at the expense of using the same amount to produce capital, intermediary, and finished goods for civilian benefit. A study by P. Terhal, for instance, states that "there is no doubt that for India, the real civilian opportunity costs implied by allocating scarce resources to military purposes have been substantial."[34] Terhal demonstrated in a study of the 1960–1970 period that scarce resources in foreign exchange, skilled manpower, research and development, and development funds had been substantially usurped by the defense sector. This diversion of resources, according to Terhal, had a "negative bearing" on capital formation by the central government and partly contributed to its decline from a growth rate of 2 percent of the GNP in the period 1960–1965 to 1 percent during 1966–1970.[35] Terhal's conclusion was supported in a broader study by Nicole Ball of several developing nations, including India.

It is virtually a truism that defense programs divert resources from the development program. Every rupee, foreign exchange unit, and skilled professional involved in the defense effort could have been utilized for development. The exceptions to this rule would be resources that were specially marshaled for the defense effort that might not otherwise have been tapped, foreign exchange generated by arms sales abroad, and skilled professional manpower trained within military establishments, particularly when an oversupply of such manpower exists in the civilian sector.

In India, it is clearly only in the matter of foreign exchange that defense uses resources that might otherwise have gone to development. At present, arms sales abroad of Indian-made weapons are insignificant, although since 1982 serious consideration is being given to exploring the overseas market. Defense production units such as Bharat Electronics and Mazagon Docks that export civilian goods abroad tend to offset the foreign exchange cost of importing

[34] See P. Terhal, "Guns of Grain: Macro-Economic Cost of Indian Defence, 1960–70," *Economic and Political Weekly* 16. no. 49 (December 5, 1981): 1998.

[35] Ibid. The Ball study is cited in note 46 below.

weapons or components for weapons. But these exports are marginal compared to the arms import bill and merely highlight the fact that perhaps even more foreign exchange might have been earned if defense production units were entirely channeled into civilian production.

Defence Minister Y. B. Chavan estimated in 1965 the foreign exchange component of India's first five-year defense plan (1964–1969) to be about Rs. 1.1 billion a year, while the Stockholm International Peace Research Institute estimated actual expenditures at Rs. 1.25 billion.[36] This amount represented between 20 to 50 percent of the annual *excess* of imports over exports during the same period. (Between 1973 and 1983, arms imports ranged from 2.6 to 10.9 percent of total imports; or an average of 6.2 percent in the eleven-year period.[37] Moreover, Emile Benoit in his study noted that such figures only indicate imports of finished equipment and do not take into account what might have been produced for export with defense funds, thereby reducing the annual trade deficit. Subrahmanyam argued in the late 1960s that because much of Indian military imports are from the Soviet Union (which allows India long-term credit at low interest rates and whose imports are offset by large quantities of Indian exports to the Soviet Union that may not have taken place otherwise), the foreign exchange burden is much less than what it may seem. But Benoit's case here was that Indian exports to Russia might well have been exported to other countries for payment in hard currencies. Indeed, Benoit points out reports that the Soviet Union and some East European countries have re-exported certain Indian goods such as railroad rolling stock and cashews for higher prices in hard currencies. He further estimated that "if industrial-type resources in the Indian defense program were directed into exports or civilian import substitutes, India might gain foreign exchange equal to about half of its foreign exchange resources, nearly three-quarters of its key development imports of machinery and parts over the years."[38] Subrahmanyam's argument has also begun to appear less convincing in view of the new arms deals concluded by India in recent years with Western Europe. The total cost of the initial Jaguar agreement was estimated at $2.5

[36] Emile Benoit, *Defense and Economic Growth in Developing Countries* (Lexington, Mass.: Lexington Books, 1973), pp. 189–91.

[37] *World Military Expenditures and Arms Transfers, 1985*, U.S. Arms Control and Disarmament Agency, Washington, D.C., p. 108.

[38] Benoit, *Defense and Economic Growth*, p. 191.

billion and the Mirage agreement at $3.3 billion.[39] The facts that these two aircrafts are eventually to be produced in India and that parts may be resold to the licensing overseas company may marginally alleviate the foreign exchange burden but will not eliminate it. The Mirage-2000 deal must also be viewed in the light of the $5.7 billion loan obtained from the International Monetary Fund to resolve the heavy Indian trade imbalance caused by expensive oil imports. However, much of this loan may be absorbed by the escalating arms import bill.

Whatever the strength of the arguments regarding the opportunity costs of defense, there can be little doubt that the opportunity costs represented by foreign exchange spending on defense are crucial. Major development projects in India require critical foreign components and foreign exchange is in perpetual short supply. Concessional credit terms provided by the Soviet Union, "buy-back" arrangements in coproduction agreements, and the export of civilian goods manufactured by defense undertakings do not resolve the problem of depriving development programs of scarce foreign exchange.

In contrast to the strong evidence of the opportunity costs of foreign exchange utilization, the weakest argument concerns the alternative use of skilled manpower. It is true that the military-industrial complex in India employs a substantial pool of the professional scientific and managerial cadres in India. These include nine public sector undertakings and several ordnance factories under the Ministry of Defence, the Department of Atomic Energy, and the Department of Space. The production of the defense ordnance factories in 1982–1983 was worth Rs. 869 crores ($869 million) and employed about 176,000 persons.[40] The public sector defense undertakings produced Rs. 1,144 crores ($1.4 billion) worth of goods in 1982–1983 and employed about 103,000 people. (See table 6.3.) The bulk of the employment here ranges from highly skilled workers to an elite corps of scientists, engineers, and managers that may number up to 20,000. The professional class tends to be even more predominant and concentrated in the Departments of Atomic Energy and Space. In 1981–1982, the Department of Atomic Energy employed 19,158 scientific and technical personnel and 12,498 administrative and auxiliary staff; and the Department of Space em-

[39] For cost estimates and assessments of the Jaguar deal, see *Times of India*, October 7, 22, and 23, 1978; for the Mirage-2000 deal, see *Times of India*, October 27, 1981, and February 18, 1982.

[40] See *Report: 1983–84*, Ministry of Defence, pp. 34–37, 44–49.

TABLE 6 3
Employment in Defense Public Sector, 1982–1983

Enterprise	*Number of Employees*	*Scheduled Castes*	*Scheduled Tribes*
HAL	40,725	5,007	975
BEL	18,072	2,918	50
BEML	13,531	2,417	289
MDL	13,741	1,113	276
GRSE	10,785	1,342	439
GSL	1,821	165	2
PTL	1,981	329	10
BDL	1,193	214	27
MDNL	1,359	202	45
Totals	103,208	13,707	2,113

SOURCE *Report, 1983–84* (New Delhi. Ministry of Defence, Government of India), p. 49
NOTE For enterprise names, see table 6 1 Employment of underprivileged former "Untouchables" and tribals is under special "schedules" of the Indian Constitution

ployed 7,136 and 4,172 such personnel in the same categories.[41] In addition to this work force, there are 1.5 million men in the armed services with an almost equivalent number of civilians in supporting capacities.[42] Of the men in uniform, about 10 percent are officers who have had graduate-level education and are trained further in command roles. They are capable of assuming managerial positions in the civilian sector.

These figures may suggest an imposing diversion of manpower to the defense sector at the cost of the development program. Yet such an assessment is misleading. The recruitment of officers and men for the armed forces and the employment of professional cadres and skilled workers for the defense production units must be viewed in the context of the abundance of manpower in India, where masses of people are unemployed or underemployed. The industrial sector of India employs only about 10 percent of total work force (about 27 million in 1978).[43] Every year the industrial sector is only able to absorb about 12 percent of those newly enter-

[41] See *Report: 1980–81* (New Delhi: Dept. of Atomic Energy, Government of India), p 5, and *Report: 1980–81* (New Delhi: Dept. of Space, Government of India), p. 37.

[42] See S. S. Khera, *India's Defence Problems* (Bombay: Orient Longmans, 1968), p. 55.

[43] Krishna, "The Economic Development of India," p. 169.

ing the work force so that there is a perpetually high level of unemployment and underemployment among India's youth. This remains true despite the high level of absorption by the defense sector. Moreover, a substantial number of the recruits into the rank and file of the military services come from the agricultural sector, where about 75 percent of the total Indian work force is located. Even here the diversion to the military cannot be considered harmful; there are high levels of unemployment or underemployment in the agricultural sector as well.

Similarly, the employment of professional cadres in the defense sector must again be viewed in the context of an abundance of such resources in India. After the superpowers, India has the third largest pool of scientists and engineers in the world, much of which remains underutilized. The annual growth of these cadres well exceeds the ability of the national economy to use them effectively. As a consequence, there is a great deal of pressure in the Indian scientific and engineering community to emigrate to the industrialized West. Indeed, there has been a constant "brain drain" of some of the best Indian scientists and engineers. To the extent that the defense sector is able to employ their talents in its various production units, it curtails the losses to the West. And since more than half of the production of the defense public sector undertakings consists of civilian goods, defense manpower contributes substantial benefits to the civilian sector.

Finally, it should be noted that within the military sector itself (as opposed to civilian-run defense production units), the recruitment and training of personnel constitute an "addition" to the professional and skilled manpower of India. Such personnel receive training in a variety of engineering fields, particularly in the specialized maintenance wings of the services—aeronautical, automotive, electronic, civil, mechanical, and electrical engineering. Since retirement in the services is usually early (between ages 35 and 45), the military releases into the civilian work force substantial numbers of qualified personnel that may not otherwise have existed. In fact, the problem is not so much the opportunity cost of professional and skilled manpower as much as the training and release of such cadres into an already oversupplied labor market. The problem is essentially one of effectively absorbing such manpower into the civilian economy once they have completed their service in the military.

Although the diversion of foreign exchange to the defense sector appears to produce a net loss for the development program and that of skilled manpower a net benefit, the opportunity cost of domestic capital is ambiguous. As in the case of foreign exchange, this argument suggests that domestic resources spent on defense could well have been spent on development. On the other hand, those who believe in the complementarity of defense and development point out that this view fails to take into account several beneficial effects: in a developing country like India, defense programs tend to tap idle resources; defense spending increases aggregate demand and generates multiplier effects, thereby stimulating the civilian economy; and defense production provides economies of scale in the automotive, shipbuilding, aeronautical, and electronics industries. Eventually, if steady defense allocations and growth are assumed, defense becomes integrated with the civilian economy and provides net benefits rather than losses.

Whether defense and development tend to be competitive or coextensive with each other remains a controversial issue. Analysts such as Subrahmanyam, Benoit, and David K. Whynes believe that military expenditures, especially in a large country such as India, may have a positive impact on the development program.[44] In Benoit's empirical study, he discovered that in 1963 and 1964, the immediate years after the Sino-Indian war when Indian defense expenditures reached highs of 4.5 and 3.8 percent of the GNP, the GNP increased at an annual rate of 6.3 percent. This compared to a 4.5 percent average economic growth rate in the period between 1950 and 1961, when defense received annual allotments of about 2 percent of the GNP. Moreover, this happened despite the fact that agricultural output rose at only 2.7 percent per annum in 1963 and 1964, thus eliminating the possibility that the spurt in economic growth may have been due to favorable monsoon rains. Additionally, Benoit's study indicated that the sudden increase in defense expenditures after 1962 did not take place at the expense of investment. The average investment ratio (that is, investment relative to the GNP) was about 16.5 percent during the three years from 1962 to 1964. This compared to an average of only 12.3 percent for the period 1951–1964. Benoit, therefore, concluded that "while this historical experience could be coincidental—and it is certainly too brief to provide any hard proof—it does tend to support the possibility that rising defense expenditures may have been, on balance, favorable to growth in India."[45] Similar findings were evident in al-

[44] See K. Subrahmanyam, *Defence and Development*; and David K. Whynes, *The Economics of Third World Military Expenditures* (London: Macmillan Press, 1979).

[45] See Benoit, *Defense and Economic Growth*, pp. 162–64.

most all the forty-four countries studied by Benoit, and the basic correlation seemed strong enough "so that there was less than one chance in a thousand that it could have occurred by accident."

Benoit also found that only a small part of an income not spent on defense was put into highly productive investments in the developing nations, although in theory direct channeling of defense funds into civilian production might have produced much better results.[46] In reality, most income went into consumption and the rest into social investments such as housing and social welfare benefits rather than investments that produce net capital formation and an increase in future civilian production. On the other side, defense programs had indirectly or inadvertently contributed to the civilian economy through investments in roads in border areas, electronic communications, and spinoffs from shipbuilding, aircraft, and vehicle production.

Analysts such as Nicole Ball, P. Terhal, and Ved P. Gandhi disagree with these assessments. Nicole Ball, concluded that Benoit's study has "a number of serious analytic shortcomings, which are related to the availability of statistics, the methodology employed, and the kinds of information Benoit chose to include or exclude from his regression analysis."[47] Ball claimed that even if Benoit's conclusions were valid, the fact remains that high economic growth rates in no way guarantee that socioeconomic development has occurred. According to Ball, developing countries with high defense spending and high economic growth rates have usually been characterized by a high incidence of poverty. The fruits of growth tend to be concentrated with little evidence of the "trickle down" effects promised by the proponents of this theory.

In the Indian case, there are difficulties in accepting either claim. As Benoit demonstrated, empirically there is little or no inverse correlation between defense spending and economic growth rates, even after adjusting for time lags. On the other hand, there is little to indicate that less defense spending is likely to be harmful to the economy. The argument that defense spending stimulates the economy through an increase in the aggregate demand is misleading, because the same results may be obtained through civilian public works programs. Benoit's case studies of Iran, Pakistan, Taiwan, and South Korea, where he demonstrates a positive correlation,

[46] Ibid., pp. xix–xxi.

[47] Nicole Ball, *The Military in the Development Process* (Claremont, Calif.: Regina Books, 1981), pp. 7–8.

must be viewed in the context of generous economic and military aid provided by the United States and liberal domestic economic policies that encouraged private investment and growth—conditions and policies that have not been part of the Indian experience. Thus, conclusions indicating a negative or positive correlation between defense spending and economic growth rates—especially at modest levels of defense spending as in the case of India—are difficult to sustain. This conclusion suggests that within limits, decisions regarding the level of defense spending in India are better determined by the criteria of strategic necessities rather than domestic economic effects on development programs.

Defense and Inflation. Similar arguments may be advanced about the inflationary effects of defense spending on the economy. On the one side, the opponents of high military expenditure argue that military programs tend to be inflationary. According to this analysis, unlike most other state services and government-produced capital and consumer goods, no real consumer of the machines and materials of war exists. Postal and railway services, for instance, are paid for by the citizens at large when they avail themselves of these services. Government-produced industrial and agricultural goods are sold to other private and public corporations and individuals for a price to cover at least the costs of production. Weapons, on the other hand, whether produced in the private or public sector, are paid for by the state. Since there is no alternative civilian use for such machines of war, they are wasted out through obsolescence or are used in war. Likewise, the armed forces provide no direct and continuous services to the economy but are paid wages and salaries that will be used to purchase other goods and services in the country.

Military expenditures are also considered inflationary because of the rapid rate of obsolescence of weapons and the need for constant change to meet changing threat conditions. Particularly in the case of India, the location of defense in the public sector implies unnecessarily large bureaucracies and massive wage bills that may further aggravate inflation. Private defense contractors and subcontractors are also more likely to quote higher prices when negotiating with the government, which is not run on a profit-and-loss basis and pays less attention to optimizing cost and quality.[48]

[48] See statement by Admiral H. G. Rickover, in *Hearings Before the Joint Economic Committee, Congress of the United States*, 97th Cong., 2d sess., pt. 1, January 28, 1982 (Washington, D.C.: U.S. Government Printing Office, 1982), pp. 9–10. See also Sivard, *World Military and Social Expenditures, 1981*, p. 18.

The inflationary potential of defense spending should be evident. The mitigation or even elimination of such inflationary consequences is usually attempted by resorting to various types of tax structures, defense bond issues, and other fiscal measures that attempt to absorb any excess demand generated by defense-related incomes and purchases. However, in the case of comparatively higher levels of defense expenditures, the ability of governments to reduce inflationary spirals has not always been successful. War and postwar economies have traditionally been known to be inflationary; the adverse effects of runaway movements tend to discourage savings and investments and, therefore, can disorganize economic planning, induce excessive stockpiling, and upset future production schedules. In the case of a developing nation such as India, economic disarray produced by inflation can also have disastrous political effects, because the majority of the population lives at a subsistence level and on a fixed or relatively inelastic income.

Until the 1962 Sino-Indian war there was little or no inflation in India. Between 1950 and 1958 there was virtually no net gain in the wholesale price index and only a net rise of 7 percent in the next four years.[49] With the defense buildup beginning in 1963, prices rose cumulatively by more than 50 percent in the six-year period from 1963 to 1968. There was a 1 percent decline during the mild depression year of 1968–1969, after which the wholesale price index registered a moderate increase of about 9 percent during the next two years, ending in 1971. Thereafter, the most serious inflation occurred between 1973 and 1975 when prices rose by more than 50 percent in the three-year period (see table 6.4). Inflation was curbed drastically during the following four years when it registered 3.9 percent in 1975–1976, −2.0 percent in 1976–1977, 5.2 percent in 1977–1978, and no change in 1978–1979.[50] It then rose 17.1 percent in 1979 and 18.1 percent in 1980 and subsequently dropped to 10.7 percent in 1981 and 5.2 percent in 1982. In 1983 and 1984 the inflation rate was 6.2 and 9.3 percent. During the fourteen-year period 1971–1985, defense allocations varied marginally between 3 and 4 percent of the GNP.

[49] See Benoit, *Defense and Economic Growth*, pp. 170–72. For a general discussion of the problem of inflation in India, see Surjit S. Bhalla, "India's Closed Economy and World Inflation," in William R. Cline and Associates, *World Inflation and the Developing Countries* (Washington, D.C.: Brookings Institution, 1981), pp. 136–65.

[50] Figures drawn from *Times of India*, February 21, 1982; *Christian Science Monitor*; Cline, *World Inflation*, p. 137; and *Far Eastern Economic Review: Asia 1985 Yearbook*, p. 158.

TABLE 6.4
Defense, Economic Growth, and Price Inflation, 1971–1982 (in percent)

	Defense as % of GNP	*GNP Growth Rate*	*Price Change*
1971–72	3.9	1.0	3.9
1972–73	3.8	−1.0	7.8
1973–74	3.1	5.3	16.3
1974–75	3.4	1.2	28.6
1975–76	3.8	9.6	3.9
1976–77	3.6	1.3	−2.0
1977–78	3.4	8.6	5.2
1978–79	3.3	5.3	0.0
1979–80	3.5	−4.8	17.1
1980–81	3.9	7.5	18.1
1981–82	3.3	5.2	10.7
1982–83		1.8	5.2
1983–84			6.2
1984–85			9.3

SOURCES Compiled from Raju G C Thomas, *The Defence of India· A Budgetary Perspective of Strategy and Politics* (New Delhi· Macmillan Co of India, 1978, Columbia, Mo.. South Asia Books, 1978), p. 37, *Christian Science Monitor* (supplement), July 30, 1982, *India. 1981* (New Delhi. Publications Division, Ministry of Information and Broadcasting, 1981), pp 34, 166–68, *Far Eastern Economic Review· Asia 1985 Yearbook*, p 158.

It would be tempting to correlate price inflation in India with the defense buildup since 1963. However, when we examine inflation together with GNP growth rates and levels of defense expenditures, the resulting patterns are somewhat erratic. Benoit points out that between 1961 and 1963, real defense expenditures rose by almost 150 percent.[51] Allowing a year lag for its consequencs, prices rose between 1962 and 1964 at an annual rate of 7.5 percent while real gross domestic product also grew by about 6.3 percent per annum. This compared to an average annual growth rate in the eleven-year period 1951–1962 of 4.5 percent, with average defense expenditures of 2 percent and an inflation rate of 0.6 percent. These figures tend to indicate that between 1962 and 1965 the growth rate was accelerated by higher defense outlays and that the moderate price inflation appears to have aided the short period of economic prosperity. (See table 6.5.) Higher defense spending appeared at least to have no adverse effect on development.

Since 1965, there appears to be no consistent pattern between de-

[51] Benoit, *Defense and Economic Growth*, p. 172.

TABLE 6.5
Defense, Economic Growth, and Price Inflation by Decade (in percent)

	Defense as % of GNP	*GNP Growth Rate*	*Price Change*
1951–1962 (annual average)	2.0	4.5	0.6
1962–1971 (annual average)	3.6	3.7	6.9
1971–1982 (annual average)	3.5	3.5	9.1

SOURCES Compiled from Raju G. C. Thomas, *The Defense of India: A Budgetary Perspective of Strategy and Politics* (New Delhi: Macmillan Co. of India, 1978), p. 37; *Christian Science Monitor* (supplement), July 30, 1982, *India 1981* (New Delhi Publications Division, Ministry of Information and Broadcasting, 1981), pp. 34, 166–68

fense spending on the one hand and economic growth rate and price inflation on the other. In 1974–1975, when defense allocations had reached a low of 3 percent of the GNP, the GNP growth rate fell to 1.2 percent while the inflation rate rose to crisis proportions of over 25 percent. The following year, defense allocations remained approximately the same while inflation was curtailed and the GNP growth rate jumped to 9.6 percent. A similar lack of pattern may be discerned in 1979, 1980, and 1981 with inflation rates running at 17.1 percent, 18.1 percent, and 10.7 percent, while GNP growth rates fluctuated between −4.8 percent and 7.5 percent. This situation has continued in the mid-1980s.

These patterns over the last thirty years indicate that although defense spending may have contributed to price inflation since 1963, it is rarely the decisive factor determining the annual increases in prices. Moreover, if steady annual defense spending levels are perceived to contribute to price inflation, often this may be beneficial to the economy. The World Bank and other observers, for instance, believe that an inflation rate of 5 to 7 percent tends to be healthy for a country like India because it draws money into productive ventures. Note also that in India inflation tends to be a phenomenon of supply shortages in various goods rather than an excess of demand generated by income. Relative increases in productivity over income generation curtail inflation to acceptable levels, as in the years 1975–1976 and 1977–1978. Thus, when defense production stimulates civilian production through economies of scale and an increase in the aggregate demand, it leads to higher growth rates relative to the inflation rate.

On balance, the defense buildup has not seriously affected India's economic planning and development programs. Instead, economic success or failure must be attributed primarily to various economic and political policies in India ever since the ambitious "Five Year Plans" were launched in 1951; and secondarily to geographical conditions and the vagaries of nature, especially the annual monsoon rains. Resource allocations to defense have been usually prudent and generally well-tailored to India's needs despite the shocks of war in 1962, 1965, and 1971. Barring some unprecedented military catastrophe in the future (e.g., massive devastation arising from nuclear war), defense plans and programs will continue to be maintained at a comparatively low profile.

CHAPTER SEVEN

Technological Strategies and Weapons Procurement Policies

ECONOMIC constraints determine the amount of resources that can be diverted to the defense effort without adversely affecting the development program. Technological constraints, on the other hand, determine the limits of defense self-reliance in India. In turn, technological capability is dependent on general economic capability and more specifically on investments that advance Indian technical expertise. Note that although defense self-reliance is dependent on India's overall technological capability, the technological needs of the civilian and defense sectors often differ in both the quality standards and the time frame within which a machine or weapon must be produced. There is also an important political connection. The extent to which India is able to assert a substantial measure of technological independence in defense production determines New Delhi's capacity to conduct independent foreign and defense policies.

The technological aspects of Indian defense policy will be discussed here on three levels: general technological considerations, weapons procurement options, and the degree of technological self-reliance in the defense or defense-related programs of the services and the nuclear and space programs.

1. General Technological Considerations

The Nature of Indian Technological Strategy

In general, the problems of research and development in India are those that affect most of the industialized and industrializing countries of the world. These problems include the high risks that accompany heavy investments in research and development that may not yield significant technological returns. In the defense field in particular, high rates of obsolescence and the demand for quality comparable to those of potential adversaries are added pressures that the civilian sector does not face. In India, moreover, the civilian sector is protected from external competition through import and investment restrictions and high tariff barriers, a privilege the defense sector does not share. Under the pressure of rapid advances in arms technology abroad, the question of how much indigenous weapons capability the country can afford remains a key question for India's defense planners.

The question of the appropriate technology needed applies to both the defense and development sectors in India. In general, appropriate technology from the development standpoint calls attention to problems of production scales, managerial and technical skills, and materials and capital available at the local level.[1] First, the planned level and scale of production must optimize unit costs of production. Second, appropriate managerial and technical skills must be available for the absorption and effective utilization of advanced technologies that need to be transferred. Third, the technology must be conducive or adaptable to readily accessible raw, semifinished or finished materials, especially those that are available from domestic sources. Finally, the capital that needs to be raised to develop the technology must be proportionate to the economic benefits that are likely to accrue from the application of such technology.

Are the above criteria also relevant from the defense standpoint? Substantially different qualitative standards of technology may be deemed appropriate in the two contexts. The quality of technology needed for development programs may not—and need not—compare favorably with what is available or used in similar projects in advanced industrialized countries. For example, the technology and quality of Indian-made cars, trucks, locomotives, cargo ships, telephones, electronic systems, or household appliances need not compare equally with those available in Japan, the United States, or Western Europe. On the other hand, the quality of weapons needed for defense usually must be comparable to those available to India's adversaries. In recent years, some of these weapons—combat fighters, antitank missiles, radar and surveillance systems—have been among the most advanced obtainable from the international arms market. Consequently, the defense technology needed by India is determined by external rather than internal factors. The technological needs of defense often call for uneconomical small-scale levels of production, highly specialized skills that may not have much relevance for other sectors of the economy, and the commitment of

[1] See John V. Granger, *Technology and International Relations* (San Francisco: W. H. Freeman and Co., 1979), pp. 100–106.

heavy financial capital that is justified by security demands rather than by the test of economic efficiency.

Despite the differences between the needs of defense and development, the overall technological growth of the country clearly has relevance for defense capabilities. The basic technology of certain industries such as automobiles, aeronautics, electronics, shipbuilding, and structural engineering are common to both the civilian and military sectors. In the light of this, a comparative assessment of India's overall technological strategy is useful.

Four basic types of national technological policies may be discerned.[2] First, the state may pursue a broad-based strategy that attempts to achieve technological autarky. Such a strategy calls for the marshaling of considerable resources and the willingness to support scientific and technological development for almost every sector of the economy. This has been the approach of the United States and the Soviet Union and to a lesser extent of Britain and France over the last three decades. Second, the state may pursue a limited policy of scientific and technological specialization, particularly if it has certain national manpower and material resources that give it some comparative advantage over other nations. A limited degree of technological independence may be achieved through this approach. Third, the state may depend on scientific research and development undertaken abroad and import the needed technology for production at home. For almost three decades after the Second World War, Japan generally emphasized this approach. Fourth, the state may already have or may seek to acquire foreign machines and equipment for which it attempts to distill and adapt the basic technology needed. This form of "reverse engineering" was attempted by China under Mao Zedung when the country was economically and technologically isolated. The strategy continues to be used by the Soviet Union in the case of badly needed western technology to which its main access is through the legal or clandestine import of western machinery.

Needless to say, the technological policies of most nations, whether industrialized or developing, tend to be a mix of the four basic strategies. In general, given the paucity of material resources and technological skills in most developing nations, their emphasis in all likelihood is on the import of technology—provided there are

[2] The first three types described here are adapted from Robert Gilpin, "Technological Strategies and National Purpose," *Science* 169 (July 31, 1970): 441–48. The fourth strategy is suggested by John V. Granger, in Granger, *Technology and International Relations*, pp. 76-85.

no external constraints. On the other hand, for states such as Israel and South Africa that constantly face the threat of economic and technological boycotts arising from serious regional security problems, the national compulsion is to achieve technological autarky, at least to the extent this is possible.

India's technology policy has been mixed, with variations over time. The government's declared objective, expressed in occasional political manifestos, has been ultimately to achieve technological autarky. In a sense, India's declared policy emulates Britain and France, if not the United States and the Soviet Union. In practice, however, India's technological policy has sought to achieve a middle position, and this it has done with varying degrees of success. Thus, for instance, primary research and development is conducted in such areas as agriculture and birth control and in some basic capital and consumer goods industries where the cost of training, equipment, and manufacturing is not prohibitive. Such research, development, and manufacturing in the civilian sector tends to be undertaken even if cheaper and better quality goods may be obtained from abroad.

However, the criteria of defense technology is largely based on national security and is therefore influenced by conditions of political necessity rather than economic viability. The need for high quality military or military-related equipment and services—no matter what the cost—has generated research and development in India in such areas as atomic energy, outer space exploration, computers (in particular integrated circuit systems such as microprocessors), and other electronic systems that are badly needed by defense. From the civilian economic perspective, some of these programs may be difficult to justify. On other hand, high-cost defense technology has made important contributions to civilian economic development. Although many projects may have been initially justified because of their military uses, subsequently some of these programs have become justifiable from the civilian standpoint as well.

The Technological Momentum and Defense Capabilities

A good example of the interplay between development and defense technologies may be seen in India's nuclear and space programs. Although cost estimates by Indian and foreign analysts vary, the present unit cost of power generated by nuclear plants is assessed to be more than the unit cost of power generated by hydroelectric and coal-fired thermal plants. Nevertheless, civilian justification often provided for developing the nuclear energy program usually rests on the need to fill in critical shortfalls in total energy requirements and the prospect that this form of energy may prove to be commercially viable in the future. However, as progress is made on the civilian front, pressures to exercise India's nuclear weapons option also increases. Similarly, the Indian space program at present may appear to rank low in India's development priorities. But here too there are civilian contributions in meteorological forecasting and telecommunications, whatever the cost, and the technological momentum generated makes it difficult to restrain or reverse the continuation of the space program. This consequently provides India with a potential nuclear weapons delivery system.[3]

Apart from the pressures arising from the needs of defense and development, the technological momentum in India arises from the autonomous growth of scientific and technical personnel in India. As discussed in the last chapter, India now has the third largest pool of scientists and engineers in the world. The growth of nuclear and space scientists and engineers has been particularly spectacular in number and quality. The Tata Institute of Fundamental Research and the Bhabha Atomic Research Center in Bombay, the Saha Institute of Nuclear Physics in Calcutta, and the Physical Research Laboratory are staffed by world-class nuclear and space scientists and engineers.[4] If these experts are not utilized by India, the probability is that they will emigrate to the industrialized West in search of employment. Many have been trained at considerable cost at the five Indian Institutes of Technology at Bombay, Delhi, Madras, Kanpur, and Kharaghpur, and at the Indian Institute of Science at Bangalore. Although graduates of these institutes are absorbed to a certain extent by Indian industry, this is not true of the nuclear and space training centers. The elimination of the government-run nuclear and space programs could mean the total loss of this group of highly qualified personnel to the industrial West.

Despite the general thirst for scientific and technological education and training in India, the government has been aware of the costly risks of investments in research and development programs. This has produced a cautious government policy in allocations to R

[3] See Maurice Eisenstein, "Third World Missiles and Nuclear Proliferation," *Washington Quarterly* 5, no. 3 (Summer 1982): 112–15.

[4] See *Performance Budget of the Department of Atomic Energy, 1980–81* (New Delhi: Dept. of Atomic Energy, Government of India, 1981), pp. 3, 9-12; and *India: 1981* (New Delhi: Publications Division, Ministry of Information and Broadcasting, 1981), pp. 73–79.

TABLE 7.1
Expenditure on Scientific Research and Development (in Rs. crores)

	1980–81	1981–82	1982–83	1983–84 (estimated)
Central government	617.22	765.15	963.56	1,108.09
State government	75.73	91.00	112.62	126.13
Private sector	120.69	147.00	161.38	193.65
Total R & D	813.64	1,003.15	1,237.56	1,427.87

SOURCE *India: 1984* (New Delhi: Publications Division, Ministry of Information and Broadcasting, 1984), p. 86

NOTE Data is given as reported in source.

& D, with implications for the goal of defense self-reliance as well. To be sure, there has been a modest but relatively steady increase in the allocations to research and develoment. The 1966 Education Commission, for instance, reported that in India the average annual allocation to R & D before 1960 was about 0.1 percent of the GNP compared to 2.8 percent in the United States, 2.3 percent in the Soviet Union, 2.7 percent in Great Britain, 0.7 percent in China, and 0.1 percent in Pakistan.[5] In 1966, the Indian allocation to R & D had increased to 0.3 percent of the GNP and by 1980 had reached up to 0.7 percent, but this was still proportionately much less than allocations in the United States, the Soviet Union, or China. (See tables 7.1 and 7.2.)

On the defense side, only 1 percent of the defense budget was allocated to R & D during much of the 1960s. By 1977, this had risen to 2 percent and by the 1980s to almost 3 percent.[6] (See table 7.3.) However, these figures are somewhat misleading because they do not include R & D in the Departments of Atomic Energy and Space, which carry a great deal of relevance for future defense programs. The combined R & D allocations for Defence, Atomic Energy, and Space constitute almost 7 percent of the total defense budget.

These trends indicate a substantial improvement in India's ef-

[5] See *Estimates Committee, 1967–68*, 12th Report, "Defence Research and Development Organization," 4th Lok Sabha (New Delhi: Ministry of Defence, Government of India, 1967), pp. 3–6.

[6] Figures derived or assessed from *Institute for Defence Studies and Analyses Journal* 4, no. 4 (April 1972):442; *The Chanakya Defence Annual, 1973* (Allahabad: Chanakya Publishing House, 1977), p. 162; and *India: 1981*, pp. 34 and 69; and *India: 1984* (New Delhi: Publications Division, Ministry of Information and Broadcasting, 1984), p. 86.

TABLE 7.2
Science and Technology Budgets of Major Central Government Organizations (in Rs. crores)

Organization	1980–81	1981–82	1982–83
Department of Atomic Energy (DAE)	73.48	88.26	105.63
Department of Space (DOS)	56.02	75.03	86.05
Defence Research & Development Organisation (DRDO)	79.70	104.83	121.00
Total of Defense and Defense-Related Research & Development	209.20	268.12	312.68
Department of Science and Technology	40.64	46.08	60.51
Council of Scientific and Industrial Research	69.00	78.77	100.81
Indian Council of Agricultural Research	97.45	111.50	131.22
Indian Council of Medical Research	9.00	11.95	14.23
Department of Electronics	5.41	7.51	8.48
Department of Environment	3.74	5.47	9.05
Department of Ocean Development		7.88	49.36
Department of Non-Conventional Energy Sources	4.00	9.90	12.03
Total of Primarily Civilian-Related Research & Development	229.24	279.06	385.69
Total of All Central Government Research & Development	438.44	547.18	698.37

SOURCE *India: 1984* (New Delhi: Publications Division, Ministry of Information and Broadcasting, 1984), p. 86.

forts at technological self-sufficiency in general and defense self-reliance in particular. However, despite the increase in defense R & D allocations over the last twenty years, the essential technological strategy has been to purchase the needed technical know-how from abroad. This more pragmatic approach is represented in both the civilian and the defense sectors. The emphasis on buying overseas technology through joint industrial production ventures is expected to pay greater dividends for the goals of self-reliance as it enables the government to concentrate larger R & D allocations in specific areas where considerable technological benefits are likely to be obtained. Thus, rather than attempt to design and develop entire weapons systems, this approach involves the acquisition of technology from abroad for larger and more sophisticated weapon systems and focuses on the indigenous design and development of smaller weapons where the risks of low technological payoffs are less. Such a policy is also expected to avoid excessive time delays and the prospect that the weapons obtained may become obsolete by the time they are produced on a large scale for the military services.

TABLE 7.3
Research and Development Allocations of Defense-Related Government Organizations as Percent of GNP and National Expenditures, 1982–1983

	Total[a] R & D (698)	*DRDO[a]* (121)	*DAE[a]* (105)	*DOS[a]* (86)	*DRDO + DAE + DOS* (312)
		(Above figures in Rs. crs.)			
GNP at current prices (Rs. 145,141 crs.)	0.48	0.08	0.07	0.06	0.23
Central government expenditure (Rs. 19,345 crs.)[b]	3.61	0.63	0.54	0.45	1.62
Defense expenditure budget estimates (Rs. 4,704 crs.)[c]	14.84	2.57	2.23	1.82	6.62
Total government R & D (Rs. 698 crs.)		17.34	15.04	12.32	44.70

SOURCE: Compiled from *India. 1984* (New Delhi. Publications Division, Ministry of Information and Broadcasting), pp. 86, 191–208, Lok Sabha Debates, 7th ser., vol. 26, no. 25 (March 25, 1982), cols. 377–40.

[a] Total R & D–research and development in all government organisations, DRDO–Defence Research and Development Organisation, DAE–Department of Atomic Energy; DOS–Department of Space.

[b] Central government expenditure includes only revenue expenditure. Disbursements on capital account amounted to Rs. 14,628 crs. in 1982–1983

[c] Defense expenditure includes both revenue and capital expenditures.

The development of Indian scientific and technological capabilities relevant for defense spans the more obvious agencies like the Defence Research and Development Organisation and the Departments of Atomic Energy and Space. More indirectly, it includes work undertaken by the Council for Scientific and Industrial Research, the Department of Science and Technology, and the Department of Electronics.

Defense Research and Development Organisation (DRDO). The functions of the DRDO under the Ministry of Defence are twofold: to conduct research and development programs at its various laboratory complexes; and to identify, promote, and coordinate the other R & D programs of the public sector both within and outside of the Ministry of Defence as well as relevant activities conducted in the private industrial sector.[7] The DRDO also maintains links with research centers of higher learning in the country.

Within the DRDO establishments, the tasks undertaken are of two types: projects that are directly in response to specific requests made by the armed services for weapons and equipment; and projects

[7] See *Report: 1979–80* (Ministry of Defence, Government of India), pp. 45–57; and *India: 1981*, p. 89.

that carry long-range relevance for security and defense technological capabilities. The scope of the DRDO's research activities and accomplishments has thus far included armaments; explosives and propellants, rockets and missiles, including inertial guidance systems, solid and liquid propulsion systems, and air-to-surface and surface-to-air missiles; electronic systems, especially radar and communication systems; aeronautics, especially gas turbine engines and avionics, simulators for training pilots, and modifications for helicopters; naval science, especially the physical, chemical, and metallurgical aspects of underwater weapons, marine corrosion, and biodeterioration technologies; structural and civil engineering; vehicles; general stores; and materials technology and applied sciences, especially those dealing with alloy armor steel, super alloys, ceramics, and chemicals.[8]

Major programs under the DRDO have included the development of an indigenous tank (presently referred to as the Main Battle Tank or MBT) that is expected to incorporate the most recent tank technology. The first prototype of the 1,500 horsepower air-cooled engine for the tank was completed in 1982. Progress has also been made on increasing the Vijayanta tank's armor-piercing capability and correspondingly, efforts are under way to improve the resistance of Vijayanta tank armor to the pounding of enemy guns. A prototype gas turbine aero-engine (presently named the GTX engine) for a future Indian-made jet fighter is expected to be completed by the late 1980s. An airborne radar to enhance air defense is under development. For the navy, the DRDO is developing technologies in sonar equipment, underwater weapons, and other electronics and materials for nautical use.

There is substantial coordination between DRDO establishments and the public sector undertakings under the Ministry of Defence, most of which have their own in-house research divisions. For example, Defence Metallurgical Research Laboratory, a division of DRDO, collaborated with Hindustan Aeronautics to produce brake pads for the MiG-21 aircraft. Other efforts include the development of sonar equipment in collaboration with Bharat Electronics and airborne fire control radar with Hindustan Aeronautics.

Efforts to increase the DRDO's interaction with civilian industries in both the nondefense governmental sector and the private sector have been initiated in recent years. The new approach was begun around 1978 when representatives of selected industrial units in the

[8] From *Report: 1979–80*, Ministry of Defence, pp. 51–57.

field of electronics were brought in for discussions with DRDO personnel. The outcome of these deliberations was to distribute defense development projects to private industries where substantial R & D capabilities exist. This revised policy was undertaken to enable DRDO laboratories to concentrate on programs of sensitive systems of a multidisciplinary nature.

The DRDO also maintains ties with universities and other educational centers to obtain the benefits of their defense-related research. At the same time, a number of overflow DRDO projects are being farmed out to these civilian research centers, usually projects that have both defense and civilian uses. The findings are often released to private industry for commercial exploitation.

Departments of Atomic Energy and Space. These two departments carry considerable relevance for Indian defense despite their officially designated peaceful purposes. As discussed above, the dual civilian-military purpose of these departments rests on their ability to provide India with nuclear weapons and delivery capabilities.[9] (See table 7.4.)

The Department of Atomic Energy was set up in 1954 as the executive agency of the Indian Atomic Energy Commission, which had been established six years earlier. Much of the research of this department is undertaken at the Bhabha Atomic Research Centre (BARC) at Trombay, a suburb of Bombay. BARC is the largest scientific establishment in the country and its annual budget equals or exceeds the combined budget of all the DRDO complexes. For instance, in the budget estimates for 1980–1981, Atomic Energy received Rs. 74 crores compared to Rs. 56 crores for the Department of Space and Rs. 80 crores for DRDO.[10] In the budget estimates for 1982–1983, Atomic Energy's share had risen to Rs. 105 crores compared to Rs. 86 crores for Space and Rs. 121 crores for DRDO.

BARC houses four research reactors: Apsara, a 400-kilowatt swimming pool type reactor; Cirus, a 40-megawatt reactor; Zerlina, a

[9] On the civilian-military motives underlying India's nuclear and space programs, see Onkar Marwah, "India's Nuclear and Space Programs: Intent and Policy," *International Security* 2, no. 2 (Fall 1977):96–121; R. Ramana, "Development of Nuclear Energy in India, 1947-73," in *Perspectives of India's Nuclear Policy*, ed. T. T. Poulose (New Delhi: Young Asia Publications, 1978), pp. 1–15; T. T. Poulouse, *Nuclear Proliferation and the Third World* (New Delhi: ABC Publishing House, 1982); Ashok Kapur, *India's Nuclear Option: Atomic Diplomacy and Decision-Making* (New York: Praeger Publishers, 1976); and Raju G. C. Thomas, "Energy Politics and Indian Security," *Pacific Affairs* 55, no. 1 (Spring 1982):32–53.

[10] *India: 1984*, p. 86.

TABLE 7.4
Atomic Energy Program Structure

R & D PROGRAM
- Bhabha Atomic Research Centre, Bombay
- Variable Energy Cyclotron, Calcutta
- Gauribidnur Seismic Station
- Reactor Research Centre, Kalpakkam
- Research Facilities in Srinagar and Gulmarg
- Aid to autonomous institutions:
  - — Tata Institute of Fundamental Research, Bombay
  - — Saha Institute of Nuclear Physics, Calcutta
  - — Tata Memorial Centre, Bombay
- Financial assistance to universities, research institutions, and for symposia/seminars

NUCLEAR POWER PROGRAM
- Atomic Minerals Division, Hyderabad
- Uranium Corporation of India Ltd., Jaduguda
- Nuclear Fuel Complex, Hyderabad
- Heavy water projects, Baroda, Kota, Talcher and Tuticorin
- Power Projects Engineering Division, Bombay:
  - — Tarapur Atomic Power Station Unit I
  - — Tarapur Atomic Power Station Unit II
  - — Rajasthan Atomic Power Station Unit I
  - — Rajasthan Atomic Power Station Unit II
  - — Madras Atomic Power Projects I and II
  - — Narora Atomic Power Projects I and II
- Fuel Reprocessing Plant, Tarapur
- Waste Management Facilities, Tarapur and Kalpakkam

INDUSTRY and EXTENSION PROGRAM
- Electronics Corporation of India Ltd., Hyderabad
- Indian Rare Earths Ltd., Bombay

SUPPORT SERVICES
- Directorate of Purchase and Stores, Bombay
- Civil Engineering Units, Bombay and Kalpakkam
- General Services Organisation, Kalpakkam
- Directorate of Estate Management and Housing Projects
- Contributory Health Services Scheme

DIRECTION and ADMINISTRATION
- Secretariat, Bombay and Delhi
- Computer and Management Services

SOURCE *Performance Budget of the Department of Atomic Energy, 1980–81*, Department of Atomic Energy, Government of India, New Delhi, p. 34.

zero-energy experimental reactor; and Purnima, a zero-energy fast reactor.[11] A 14-megawatt fast breeder reactor and a 100-megawatt thermal research reactor (the Dhruva reactor)—both built indigenously—became operational in 1985. BARC's activities encompass the development of over 350 radioactive products that are supplied to some five hundred institutions in India and abroad. These range from radioisotopes used in industrial radiography to the diagnosis and treatment of diseases. Additionally, BARC has also set up a Variable Energy Cyclotron in Calcutta for advanced research in nuclear physics; a seismic station near Bangalore for the detection and identification of earthquakes and underground nuclear explosions; and a High Altitude Research Laboratory at Gulmarg for research on elevated conditions.

In addition to research and development, the Atomic Mineral Division of the Atomic Energy Commission is engaged in surveying, prospecting, and developing uranium, thorium, beryllium, and columbium tantalum. Another subdivision, Indian Rare Earths Ltd., extracts rare earths, minerals, and thorium from the beach sands of South India and Orissa. The Uranium Corporation of India mines and concentrates uranium in Bihar. Finally, a nuclear fuel complex at Hyderabad fabricates fuel elements for nuclear power reactors.

Space exploration began in 1962 when a committee for space research was established under the auspices of the Department of Atomic Energy. Seven years later the Indian Space Research Organisation (ISRO) was created with headquarters in Bangalore to conduct research and activities in space science and technology. In 1972, a Space Commission and a Department of Space, separate from the Department of Atomic Energy, were set up alongside the ISRO in Bangalore. The research activities of the ISRO are carried out at five space centers: the Vikram Sarabhai Space Center (VSSC) in Trivandrum, Kerala; the ISRO Satellite Center (ISAC) in Bangalore, Karnataka; the SHAR Center at Sriharikota, Andhra Pradesh; the Space Applications Center (SAC) at Ahmedabad, Gujarat; and the National Remote Sensing Agency (NRSA) at Secuderabad, Andhra Pradesh. The major center for research and development in space science is the Physical Research Laboratory in Ahmedabad, a facility that is supported by the Department of Space.

Other Defense-Related Agencies. Like the Departments of Atomic Energy and Space, the Electronics Commission and the Department

[11] See *Performance Budget of the Department of Atomic Energy, 1980–81*, pp. 9–12.

of Electronics sponsor research and development that have implications for defense programs.[12] The Department of Electronics coordinates the R & D of electronics industries in the public and private sectors. In the public sector, the production units include Bharat Electronics and Hindustan Aeronautics (both under the Minstry of Defence), Indian Telephone Industries, Electronics Corporation of India (the major producer of indigenous computers), Instrumentations Ltd., Hindustan Teleprinters, and Central Electronics Ltd. These undertakings account for more than 40 percent of total electronics production in India.

In order to establish a broad-based and long-term technological base, the Electronics Commission has promoted research and development projects in various research institutions, universities, and the five Institutes of Technology. Projects supported by the commission include research on radars, sonars, and navigational aids, and other sectors of electronics such as materials, components, instruments, controls and computers. In particular, the Department of Electronics undertakes research on electronic computer systems for the Ministry of Defence.

Two remaining broad-based R & D departments are the Department of Science and Technology and the Council of Scientific and Industrial Research (CSIR). The former promotes pioneering research in the country, the latter's purpose is to organize, coordinate, and optimize the utilization of the large pool of scientists in India. Finally, it should be noted that the various R & D organizations in India are monitored by an apex body called the Cabinet Committee on Science and Technology, which is headed by the prime minister. This committee reviews the progress of the overall science and technology programs and makes decisions on high level policy matters, corrective measures, and major investments.

II. Weapons Procurement Policies

Policies of weapons acquisition in India are colored by the objectives of those who make the decisions. From the political standpoint, the primary goal is to assert the nation's independence in defense policy making, and this is best achieved through technological self-reliance. From the military standpoint, the most important objective is to obtain weapons that are comparable technologically to those obtained by potential adversaries, and this is best achieved by gaining

[12] This general information is drawn from *India: 1981*, pp. 68–92.

access to the latest weapons available overseas. From the economic standpoint, the need is to strike an optimum balance between the cost and quality of weaponry; this is best obtained in the long run through both external technological transfers and domestic production. Thus India's politicians prefer technological independence, its military prefers qualitative sufficiency, and its economic planners prefer to settle for something between the two. Implicit in all of these choices is the question of India's technological capabilities, which ultimately determine which preferences are feasible. Issues underlying India's weapons procurement policies may be assessed by examining three basic strategies—broadly categorized as overseas purchases, indigenous production, and licensed production in collaboration with an overseas manufacturer—that have been adopted over the last thirty years.

Overseas Purchases

The principal advantage of buying weapons in the international arms market is that it means the deployment of the latest weapons being produced by the advanced industrialized nations. This policy may appear imperative if India's adversaries are able to obtain similarly advanced weapons. Moreover, the policy would be inevitable if India could not produce the needed weapons of high quality on its own. A further advantage of direct overseas purchases is the ability to determine current needs according to the prevailing weapons technology and its availability in the market. Thus the government may choose to embark on a defense modernization program in response to enemy acquisitions at any time by entering the international arms market rather than having to plan several years in advance if weapons are to be obtained indigenously or through licensed production.[13]

Most of these pressures were evident in the 1950s and to a lesser extent in the 1980s. Immediately after Independence, India lacked the economic and technological infrastructure necessary to embark on a major weapons production program to meet its defense needs.[14] At the same time, New Delhi was faced with a major Paki-

[13] For discussions on this topic, see P. R. Chari, "Defence Production in India: Import or Manufacture?" *Strategic Analysis* (New Delhi) (February 1978):1–6; K. R. Singh, "Progress of Defence Production," *Economic Times*, November 9, 1980; and K. Subrahmanyam, "Planning Defence Production," *Times of India*, January 24, 1980.

[14] For description and analyses of the arms buildup on the subcontinent before 1965, see Lorne J. Kavic, *India's Quest for Security: Defense Policies, 1945–1965* (Berkeley and Los Angeles: University of California Press, 1967); and Raju G. C. Thomas,

stani arms buildup composed of sophisticated American weapons acquired through SEATO and CENTO defense pacts. Under these circumstances, India turned to Britain and France for the purchase of an assortment of tanks and aircraft. Pakistani acquisitions of American F-104 Starfighter, F-86 Sabre, and B-57 Canberra combat aircraft and M-47/48 Patton tanks were matched by Indian purchases of British and French Hunter MK-56 FGA, B-1 Canberra, and Mystère-4A aircraft and Centurion and AMX-13 tanks.

A similar Indian response in part may be seen in the early 1980s when the United States decided to sell Pakistan the advanced F-16 fighter aircraft, M-48A tanks, and TOW antitank launchers and missiles. Although the Indian infrastructure had greatly improved by 1980, this transfer of American arms prompted India to purchase 150 Mirage-2000 aircraft, the Exocet air-to-surface missiles, and the Milan antitank guided weapons.[15]

A related but different type of experience may be seen in India's efforts to obtain American and British weapons in the aftermath of the 1962 Sino-Indian war. Initially, India's request for the transfer of the F-104 Starfighter aircraft, various artillery pieces, and naval vessels were made dependent on Indo-Pakistani resolution of the Kashmir dispute. Britain and the United States were unwilling to arm India over the objections of Pakistan.[16]

Therefore, from a long-term and overall perspective, the policy of overseas arms purchases carries several drawbacks. Accessibility to weapons comparable to those of adversaries may not be always possible because of political obstacles. When such weapons are available, there may be overt or implicit political conditions laid down in concluding the sale. Even when limitations are placed on the agreement, the government of the manufacturing country may subsequently practice the art of "spare parts diplomacy." Here the delivery of parts for the weapons sold remains under the constant threat of being withheld unless the recipient nation adjusts its foreign and defense policy to the liking of the supplier nation. Egypt's experience with the Soviet Union and Pakistan's experience with the United States are illustrative of the pitfalls of acquiring weapons directly from the superpowers.

The Defence of India: A Budgetary Perspective of Strategy and Politics (New Delhi: Macmillan Co. of India, 1978; Columbia, Mo.: South Asia Books, 1978), pp. 143–219.

[15] For the Milan deal, see *Overseas Hindustan Times*, September 17, 1981; for the Mirage-2000 agreement, see *Times of India*, April 14, 1982; and for reports of the planned purchase of the Exocets, see *Times of India*, November 29, 1982.

[16] See Thomas, *Defence of India*, pp. 45–48.

In addition, there are certain economic and technological drawbacks. Economically, such a policy implies a heavy drain on India's foreign exchange, which is crucial to the economic development program and perennially in short supply. The decisions to buy British and French weapons in the 1950s and French weapons in the 1980s have been costly affairs, involving a severe depletion of foreign exchange. Technologically, the policy implies that India will never gain the expertise it needs to achieve a good measure of self-reliance in defense.

Indigenous Production

A policy to develop and produce the needed weapons within India, assuming the ability to do so, avoids most of the drawbacks of buying arms from abroad. Indigenous production would enable India to assert an independent foreign and defense policy and would also carry considerable economic and technological benefits. Economically, it would allow the saving of valuable foreign exchange, add to the industrial production capacity of India, create new employment opportunities, and generate long-term multiplier effects on the national economy that would result from such military production and employment. Technologically, the policy would mean a substantial increase in the scientific and technical knowledge of the country with attendant benefits for the civilian sector as well.

Needless to say, indigenous weapons production is the desired option if economy and comparative quality can be obtained. Indian efforts toward these goals have been partially successful, especially in areas in which the technology is less complex and time-consuming. However, in the case of major weapons systems such as combat aircraft, tanks, and warships, the policy has displayed several weaknesses. Given the limited research and development facilities in India compared to those in advanced industrialized nations, indigenous design and development has carried with it the risk of high cost and low technological returns. When indigenous design and development have been carried forward into actual production, the process has entailed technical setbacks and excessive delays so that by the time the weapon is delivered, it is virtually obsolete. The Indian Ministry of Defence has conservatively estimated that sophisticated weapons take over twenty years from the point of design to development, production, and delivery.[17] Fewer than fifteen years are needed to manufacture a comparable weapon in India in collab-

[17] See *Report: 1979–80*, Ministry of Defence, p. 28.

oration with an overseas manufacturer. Given Indian technical limitations as well, the quality of the completely indigenous product usually fails to measure up to international standards. Moreover, the limited resources available and relatively small quantities ordered have tended to escalate the unit cost of production so that the final product costs more than an equivalent and better quality weapon that could have been imported.

Experience with the design, development, and manufacture of the HF-24 Marut fighter-bomber is illustrative of the problems.[18] In India, reconciling the strategic perspectives of the political decision makers, the aircraft specifications and standards of efficiency of the Indian air force, and the capabilities of the scientific and engineering personnel available at Hindustan Aeronautics Ltd. has proven to be difficult. Air force demands for performance have usually been much higher than HAL could cope with, and much time has been lost in arriving at a compromise. The project has required setting up facilities to train aviation scientists and engineers and the creation of sufficient and rewarding engineering opportunities to attract the high caliber of personnel required by the industry. Attempts to achieve this also carried the danger that when such a pool of engineers and scientists was in fact established, they might be tempted to go overseas in search of more lucrative opportunities.

Nevertheless, despite these setbacks, a special investigation committee headed by cabinet minister C. Subramaniam recommended in 1969 that India should persevere with the policy of designing and developing combat aircraft.[19] The committee's recommendation was based on the belief that drawbacks must be accepted during a prolonged interim phase if a pool of aeronautical scientists and engineers is to be nourished in India for the more distant goal of defense self-reliance. There is some evidence that this policy is bearing fruit at least in areas other than that of combat aircraft. Increasing success in the development and production of advanced tanks and armored personnel carriers suggests that combat aircraft design and production may eventually be successful as well. Much depends

[18] See Mahraj K. Chopra, "HF-24 and Development of Air Power," *Indian Express*, November 29, 1967; Dilip Mukerjee, "Lessons of the HF-24 Project," *Times of India*, October 24, 1970; and Thomas, *Defence of India*, pp. 193–96.

[19] For a discussion of the committee's recommendations, see Nihal Singh, "Aeronautics Plans Reflect Refreshing Realism," *Statesman*, June 6, 1969. See also article written by one of the members of the Subramaniam Committee, S. Krishnaswami, entitled "Policy for Aircraft Development," in the *Institute for Defence Studies and Analyses Journal* (New Delhi) 1, no. 2 (January 1969), pp. 75–81.

on how the development of the new gas turbine engine proceeds and the ability to match the GTX engine with an airframe of suitable standards.[20]

Licensed Production

The third policy option—the most favored in recent years—has been to seek technological collaboration with overseas manufacturers for the production in India of needed weapons.[21] In effect, this policy is a compromise between the first and second options. The production of foreign weapons under license in India provides access to modern technology from abroad while also increasing technological expertise in India. Although the weapons produced by this method may become somewhat obsolete by the time the technology is transferred and the materials and parts are obtained indigenously, this problem is not quite the same as the procurement strategy of domestic design, development, and production. With indigenous production, as noted, it may take up to twenty years for initial delivery to commence. On the other hand, the strategy of licensed production usually implies initial assembly of the weapon with imported parts and thus production and delivery usually begin within four years of the licensing agreement even if the process of indigenization takes up to fifteen.

Licensed production also allows a greater saving of foreign exchange than overseas purchase. It is true that the initial burden of foreign exchange expenditure may be quite heavy and not fundamentally different from that of buying the weapon directly, because it involves the purchase of all the major parts for assembly in India as well as the cost of royalties to be paid to the manufacturer during the licensing period. One report, for instance, suggested that the cost of producing the MiG-21 in India in the early phases may have been more costly than the direct purchase of the aircraft. Nevertheless, as indigenization progresses and royalty fees are fully paid, there is a major saving of foreign exchange. And unlike domestic design and development, the risks of investment and poor techno-

[20] See Inder Malhotra, "Beyond the Jaguar Deal: Promises and Pitfalls," *Times of India*, October 18, 1978; and "The Aeronautics Industry: Implications of the DPSA Deal," *Times of India*, September 21, 1978.

[21] For a broader discussion of technology transfers, see Prem Shankar Jha, "Free Technology Imports: Tailoring Policy to Reality," *Times of India*, October 19, 1982. For a critique of India's policy of licensed production with reference to the decision to produce the West German HDW-209 submarines, see Bharat Karnad, "Substandard Sub for the Navy," *Hindustan Times*, November 21, 1982.

logical returns are much lower under this policy because the designs obtained are usually those of proven overseas weapons. Technical setbacks are also likely to be fewer, because most such problems will already have been resolved in the licensing country; much time is saved by moving straight to the production process.

Finally, licensed production promises the possibility of political independence once the technological transfer has taken place and the materials and components are obtained and produced within the country. There is, of course, an interim phase during which the collaborating manufacturer may be compelled by his government to curtail or even terminate the transfer of technology for political reasons. This risk may seem as costly as the withholding of spare parts at a later date when weapons are purchased directly from abroad. In both cases heavy initial investment by India could be brought to nought by the supplier or the licensing country, and both policies suggest that India could be subject to external political pressures. However, if collaboration has persisted for some years, it may be possible for Indian engineers to extrapolate from already existing knowledge. The experience and performance of China after the Soviet Union pulled out its engineers and scientific advisors in the late 1950s is illustrative. China was able to develop and adapt the MiG-19 aircraft (renamed the J-6) and the T-55 tank (renamed the T-59) as well as to proceed successfully with the development of nuclear weapons. On balance, therefore, licensed production involves greater economic and technological benefits and fewer political risks than overseas purchases, although even here problems are not completely resolved.

Even in the 1950s, when the emphasis was on overseas purchases, and in the 1960s, when the rhetoric stressed the need for self-reliance, India's weapons procurement strategy has always resorted to licensed production in varying degrees. In 1957, India entered into an agreement with Folland Aircraft and Bristol-Siddley engines of Britain for the manufacture of the Gnat (now renamed the Ajeet) interceptor in India. In 1962, an agreement was reached with the Soviet Union for the manufacture of the MiG-21 interceptor. This was soon followed by a host of other licensing agreements in the 1960s: the Chieftan (renamed Vijayanta) tank in collaboration with Vickers-Armstrong of Britain; the Leander-class frigates in cooperation with Vickers and Yarrow of Britain; the Shaktiman truck with West Germany and the Nissan jeep with Japan; and the Avro-748 (renamed HS-748) with the Avro Company and Rolls Royce of Britain. There was a lull in reaching further agreements for major weapons until 1979, when an agreement was entered into with British Aerospace and Dassault of France for the initial purchase and later manufacture in India of the Jaguar fighter-bomber. In the early 1980s, there were more agreements: the manufacture of the T-72 tank and MiG-27 fighter-bomber in collaboration with the Soviet Union, the SK-1500 Type 209 submarines with Howaldtswerke Kiel (HDW) of West Germany and, tentatively, the eventual manufacture of the Mirage-2000 in India in collaboration with France. In addition to these items, a variety of sophisticated electronic, mechanical, and civil engineering components and products are made under license by public sector enterprises under the Ministry of Defence.

Review of the Options

The complexities underlying the three procurement strategies may be observed in India's nuclear and space programs. Especially in the case of the nuclear energy program, the ability to purchase sensitive equipment or even to obtain technology from abroad to conduct projects under license in India may be severely restricted or may even prove impossible because of international efforts to prevent nuclear proliferation. At the same time, the technology must be obtained from abroad, especially during the initial phases of the program.

Thus the Indian nuclear energy program began with the purchase of a research reactor, Cirus, from Canada in 1955.[22] About the same time a swimming pool-type research reactor, Aspara, was developed by Indian scientists and engineers. Subsequently, in accordance with the terms of a 1950 Indo-Canadian agreement, Canada supplied two heavy water reactors for setting up the Rajasthan power plant. A similar 1963 Indo-U.S. agreement provided for the supply of two General Electric light water reactors on a turnkey basis. With the technologies gained initially from overseas purchases and foreign collaboration, the Indian nuclear energy program has grown increasingly indigenous. The second strategy of indigenous production has become a matter of practical policy following the cutoff of Canadian and American cooperation in the aftermath of the 1974 Indian atomic test. This strategy is becoming more feasible in the nuclear sector as India's domestic scientific and technological capability continues to grow. In comparison, external pressures on

[22] See Joseph A. Yager, ed., *Nonproliferation and U.S. Foreign Policy* (Washington, D.C.: Brookings Institution, 1980), p. 95.

the Indian space program that began in 1962 are comparatively fewer because the missile delivery implications of the program appear less obvious than the weapons implications of the nuclear energy program. The program here has involved all three procurement strategies and continues to do so.

Overall, each strategy carries certain advantages and disadvantages. Although the present Indian procurement policy tends to place greater emphasis on licensed production, the question of which strategy is most preferable depends on the objectives of the policy maker. As noted earlier, for the politicians the order of preference is indigenous production, licensed production, and overseas purchase, because the essential objective is to minimize external dependence and political pressures. In contrast, the military would prefer a reverse order: overseas purchase, licensed production, and indigenous production, to maximize effective firepower and to match or overtake the enemy's military capabilities as quickly as possible.

The present order of preference, which emphasizes licensed production with what appears to be more or less equal weight placed on overseas purchases and indigenous production, is not merely an unhappy compromise between political and military desires. It represents the optimum long-term strategy of maximizing both firepower and domestic technological capability. The present emphasis also represents a shift in India's overall technological policy from the Nehru era of self-reliance to the Indira Gandhi era of self-sufficiency. Under Nehru, the belief existed that India could establish an independent technological base in the foreseeable future through a crash series of Soviet-type Five Year Plans. Under Mrs. Gandhi, the policy gradually shifted to one of acquiring domestic technological capability together with a policy of gaining assured access to overseas weapons and technology.

Underlying this shift was the rejection of complete autarky in weapons procurement and the acceptance of what one Defence Ministry official called "indigenizing Chinese-style."[23] This approach suggested that domestic technology could be nurtured and its growth accelerated by taking foreign weapons systems and adapting and improving these to India's needs. Successful Chinese efforts with the T-55 tank and MiG-19 aircraft have parallels in Indian adaptations of the Gnat and MiG-21 interceptors and the Chieftan tank. The attitude of the Indira Gandhi government also

[23] Interview with the author in 1980–1981.

stemmed from the belief that autarky, although feasible in various civilian fields of production, cannot be attempted in the defense field because of the much higher and more costly levels of technology required. Civilian consumer and heavy engineering goods need not match the quality available in the advanced industrialized countries; defense equipment, on the other hand, must equal those weapons available to India's adversaries from the industrialized nations.

III. Level of Technological Capabilities

The manner in which the three alternative procurement strategies have been applied in defense and defense-related programs is examined further in this section.

Army Defense Capabilities

The policy of indigenous development and production is most applicable to army defense programs. Weapons programs proposed by the army appear economically and technologically more viable than those of the other two services.[24] Army programs are more labor intensive in comparison with the more capital-intensive naval and air force programs. This applies both to recruitment and to the production of weapons; it is important to note in this context that the trade-off between technology and manpower is an important consideration in India in assessing the acceptability of proposed defense programs.

Particularly because an extensive technological infrastructure for this purpose has been established over the last twenty years, most of the weapons proposed by the army are capable of being designed, developed, and produced within the country. Once a decision is made to proceed with a weapon proposed by the army, there are usually fewer technical setbacks and delays in developing and producing the weapon than with the other services, because most of the army requirements have been for relatively light and simple weapons such as semiautomatic rifles, carbines, light machine guns, mortars, and ammunition. There have been comparatively fewer orders for field artillery, antiaircraft guns, and heavier machine guns.

However, some heavy equipment has not been so amenable to indigenous development and production. These include tanks, ar-

[24] See Thomas, *Defence of India*, pp. 151–55; and *Report: 1979–80*, Ministry of Defence, pp. 13–15.

TABLE 7.5
Army Weapons Procurement, 1950–1985

Indigenous Production

7.62mm Ishapore rifle, carbines, light machine guns
2" 81mm and 120mm mortars; 75/24 pack howitzers
a variety of bombs, explosives, and ammunition
106mm recoilless guns; L-70 antiaircraft guns
Main Battle Tank (prototype completed)

Licensed Production

Vijayanta tank (British)[a]
guns for the Vijayanta tank (British)[a]
Nissan jeep (Japan)[a]
Shaktiman truck (West Germany)[a]
Carl Gustav ATW (Sweden)
SS-11 ENTAC ATGW (France)[b]
T-72 tank (Soviet Union)

Overseas Purchases

Tigercat SAM (British)
SAM-6, -7, -8A, and -9 (Soviet Union)
SS-11 ENTAC ATGW (France)[c]
Milan ATGW (France/West Germany)
AMX-13 tank (France)[d]
Centurion tank (Britain)[d]
T-54/55, PT-76, and T-72 (initial batch) tanks (Soviet Union)

[a] Weapon now entirely or almost entirely indigenized.
[b] Licensed production of weapon has not yet commenced as of 1985
[c] Weapon is also to be produced under license in India.
[d] Weapon phased out or is being phased out.

mored vehicles, antitank guided missiles, and surface-to-air missiles. Although such weapons demand a higher level of technological sophistication, potentially they are capable of being developed and produced in India. As a consequence, the army's weapons procurement strategy has moved smoothly and steadily from an emphasis on overseas purchases in the 1950s and 1960s, to an emphasis on licensed production in the 1960s and 1970s, and finally to the indigenous production of most of the major categories of army weapons in the 1980s. (See table 7.5.)

The tank acquisitions of the army are a good example. In the 1950s, these consisted of purchases of the Centurion and AMX-13 tanks from Britain and France. In the 1960s, purchases continued of the T-54/55 and PT-76 tanks from the Soviet Union. At the same time, licensed production began of the Vijayanta tank (based on the design of the British Chieftain) in collaboration with Vickers-Armstrong of Britain. The technology of the tank has since been transferred and the parts and materials are now more than 95 percent indigenous. Meanwhile, design and development of an Indian-designed tank began in the 1970s and a prototype of this "Main Battle Tank" (MBT) has now been completed. However, as of 1985, there has been no production of the MBT. On the other hand, an agreement was reached in 1981 to purchase several T-72 tanks from the Soviet Union and to manufacture them subsequently under license in India. By 1985, 300 T-72 tanks had been purchased from the Soviet Union and facilities had been established for their manufacture at Avadi, where the Vijayanta is now being produced.[25] About 1,000 T-72M tanks were on order from Avadi in 1985.

A similar but unsuccessful effort has been made in recent years in the case of antitank guided weapons (ATGWs). After two decades of buying various types of this weapon—the SS-11 ENTAC and Milan ATGW from France—a move has been made toward the licensed production of the SS-11 ATGW with France and the Carl Gustav antitank weapon in collaboration with Sweden. There were efforts also to purchase initially and later to manufacture under license the American TOW antitank guided weapon in India. Efforts here failed because Washington refused to accept Indian requests for the coproduction and transfer of the technology to India after initial sales of the weapon.

Air Force Defense Capabilities

The main combat systems of the air force tend to be more expensive and technologically more complex than those of the other two services. These problems are compounded by the fact that combat aircraft tend to have higher rates of obsolescence and loss (during battle and through flying accidents in peacetime). Unlike army and naval weapons, whose battle damage often can be repaired and for which peacetime damage is rare, combat aircraft losses tend to be total and irrecoverable. Additionally, the politics underlying aircraft procurement, especially through overseas purchases and licensed production, have greater national and international news visibility, making economic and technological issues even more complex and confusing.

As in the case of the other two services, air force procurement

[25] See *The Military Balance, 1985–86* (London: International Institute for Strategic Studies, 1985) and *Report: 1983–84*, Ministry of Defence, p. 36.

TABLE 7.6
Air Force Weapons Procurement, 1950–1985

Indigenous Production
HF-24 Marut fighter-bomber[d]
HJT-16 Kiran jet trainer
Gas Turbine Experimental (GTX) advance aero-engine
various electronic equipment and avionics for aircraft
Licensed Production
Gnat/Ajeet interceptor (Britain)[a,d]
Jaguar FBA (Britain and France)
HS-748 transport (Britain)[a]
Alouette-III/Cheetah/Chetak helicopters (France)[a]
MiG-21 interceptor (various versions) (Soviet Union)[a]
MiG-27 MRCA (Soviet Union)[b]
Orpheus-701 and Orpheus-703 aero-engines for Gnat and Marut (Britain)[a]
Dart RDA-7 turbojet engine for HS-748 (Britain)[a]
Artouste-III helicopter engines for Alouette (France)[a]
Aero-engine for MiG-21 (Soviet Union)[a]
Overseas Purchase
Ouragon FGA, Mystère-IVA FGA (France)[d]
Mirage-2000 (France)
Hunter Mk-56 FGA and B-1 Canberra bomber (Britain)[b]
Jaguar FBA (Britain and France)[c]
MiG-23 FBA, MiG-25 reconnaissance, and MiG-29 (Soviet Union)
AN-12 transport,[d] AN-32 transport, and IL-76 transport (Soviet Union)
Sukhoi-7B FBA (Soviet Union)[d]
DHC-3, 46C-119G, C-119G, and C-47 transports (Union States)[d]
Viscount transport (Britain)[d]
Mi-4,[d] Mi-8, Mi-17, Mi-24, and Mi-26 helicopters (Soviet Union)
Sikorsky-62 and Bell-47G helicopters (United States)[d]

[a] Weapon now entirely or almost entirely indigenized.
[b] Licensed production of weapon has not yet commenced as of 1984.
[c] Weapon is also to be produced under license in India.
[d] Weapon phased out or is being phased out

strategies have spanned overseas purchases, licensed production, and indigenous production. (See table 7.6.) The emphasis has shifted over the last thirty years among these three, moving in sequence from purchases in the international arms market to coproduction and then to cautious attempts to develop and produce the needed aircraft in the country. As noted earlier, in the 1950s there were major purchases of Hunters, Canberras, Ouragons, and Mystères from the British and the French. The emphasis shifted in the 1960s to the coproduction of the Gnat interceptor with Britain, the

MiG-21 interceptor with the Soviet Union, the HS-748 transport plane with Britain, and the Allouette helicopter with France. Although this emphasis on licensed production continued, the official rhetoric also stressed the need for domestic development and production. Accordingly, as the Indian-made HF-24 Marut fighter-bomber began to be delivered to the Indian air force in the late sixties, research was stepped up for the development of an Indian-made aero-engine for what was expected to become India's first supersonic jet fighter with above Mach-2 performance capability.[26]

However, the arms buildup of advanced and sophisticated aircraft in some Middle East countries since 1973, accompanied by Pakistani acquisitions of the French Mirage-IIIs and -Vs and efforts (albeit unsuccessful) to obtain the A-7 Corsair fighter-bombers from the United States, escalated both the level of aircraft needed by the Indian air force and the urgency of its demands. Pakistan's later ability to obtain the advanced F-16 fighter from the United States further demonstrated the futility of attempting to develop combat aircraft within India—at least in the near future.

The primary Indian experience with the indigenous production of combat aircraft has been the design, development, and production of the HF-24 Marut fighter-bomber. Problems here highlight the difficulties of achieving self-reliance in sophisticated aircraft in a developing country like India. The design and development of this plane were initially supervised by a West German design engineer, Kurt Tank, together with other German advisors. But the Germans worked largely on their own and did not transfer much of their technical expertise to Indian engineers when they left.

To make matters more complicated, the designing of the airframe was not accompanied by the designing of a suitable aero-engine. Instead, the airframe was expected to be matched with a suitable engine to be produced under license in India in collaboration with Bristol-Siddley of Britain. At the time when the airframe was being designed, the British company had promised the development of a new powerful engine, the Orpheus-12, which was to be

[26] For a discussion of earlier air force weapons procurement, see Thomas, *Defence of India*, pp. 177–83; Raju G. C. Thomas, "Aircraft for the Indian Air Force: The Context and Implications of the Jaguar Decision," *Orbis* 24, no. 1 (Spring 1980):85–101; Bharat Karnad, "Paper Jaguar," *Overseas Hindustan Times*, August 17, 1981; Inder Malhotra, "Beyond the Mirage Deal: Future Needs of Air Defence," *Times of India*, February 18, 1982; Air Vice Marshal M. B. Naik (Retd.), "New Imperatives of Air Defence," *Times of India*, August 13, 1981; and *Report: 1979–80*, Ministry of Defence, pp. 24–27.

used also for the production of a proposed combat aircraft for some NATO countries. However, the Orpheus-12 engine failed to materialize because NATO's technological demands had escalated considerably since the late 1950s.[27] Consequently, Bristol-Siddley offered India the older and less powerful Orpheus-703 engine, which was qualitatively much lower than the air force had expected.

Efforts were then made to collaborate with Egypt to manufacture the E-300 engine and to make the HF-24 an Indo-Egyptian combat plane. But Egyptian efforts to develop and produce the E-300 engine proved unsuccessful and eventually the HF-24 airframe was matched with the Orpheus-703 engines, which were then manufactured under license in India. Thus the HF-24 Marut was only partly indigenous in design and development, and its speed and maneuverability did not meet the Indian air force's standards when it was delivered in the late 1960s.

More recently, however, Hindustan Aeronautics Ltd. has achieved some success in the development of the GTX engine, the first Indian-designed aero-engine capable of near Mach-2 speed. The cost of developing this advanced afterburner engine at the gas turbine research establishment in Bangalore was approximately Rs. 19 crores, and almost twenty years of effort had been put into it.[28] By the early 1980s, HAL proposed that further development of the plane could lead to its commercial use within a few years; the cost of this move was estimated at about Rs. 60 crores. Continued development in this area was also expected to fulfill the "entire spectrum of power plant requirements of future military aircraft" and to set the bases for a self-sustaining domestic aeronautics industry. These developments and goals appeared optimistic in 1984 and it would be premature to call the GTX engine a success. What is noteworthy is that no advanced airframe is currently being developed to match the GTX. Ironically, therefore, while the past saw the design and development of an Indian airframe but no parallel development of a suitable Indian jet engine, the future may see the successful development of an aero-engine with no parallel development of a suitable airframe.

The problems associated with indigenous design and develop-

[27] For problems associated with the HF-24 aircraft project, see *Hindu*, June 2, 1962; and Mukerjee, "Lessons of the HF-24 Project." For a parliamentary committee evaluation of the problems that beset this project, see *Public Accounts Committee, 1966–67*, 10th Report, para. 10 of Audit Report (Defence Services), entitled "Manufacture of Engines," 3d Lok Sabha, Lok Sabha Secretariat, New Delhi.

[28] See Thomas, "Aircraft for the Indian Air Force."

ment of combat aircraft in India may appear resolved by the periodic Indian efforts to purchase the weapons from abroad. However, the Indian experience here has displayed problems of a different kind. In the first place, such purchases have often tended to be near panic decisions, which result in a heavy expenditure of valuable foreign exchange and in the procurement of aircraft that are not necessarily superior to those obtained by the adversary. For example, the purchase of the Anglo-French package of Hunters, Canberras, Ouragons, and Mystères in the 1950s did not compare to the quality of the American-supplied Starfighters, Sabres, and Canberras to the Pakistan air force. The Indian decision to offset immediately Pakistani military acquisitions through SEATO and CENTO involved considerable cost but gave India no qualitative edge in the military air balance over the subcontinent. A similar reaction and result occurred in 1982 when India decided to buy 150 Mirage-2000s from France at a cost of $3.3 billion in response to the U.S. decision to sell Pakistan 40 F-16 fighters.

The Mirage-2000 purchase agreement in 1982 constituted the most costly single purchase of a weapon system by India and involved an aircraft that had not been fully developed or tested at the time by the manufacturing country. On the other hand, the F-16 has already been deployed by the U.S. air force and by some of the NATO countries. Although some vague consideration had been given earlier to the licensed manufacture in India of the Mirage-2000 and of the partly British-developed Tornado multi-role combat aircraft when the F-1 Mirage and Jaguar aircraft were also being considered, the sudden decision to sign an agreement with France for the purchase of the Mirage-2000 was triggered by the U.S.-Pakistan F-16 agreement, which was seen as introducing aircraft technology into the subcontinent that greatly superseded existing or planned Indian combat aircraft. Indeed, following the American decision to supply the F-16 aircraft to Pakistan, there have been moves to proceed with only the first two legs of the three-part Jaguar agreement with Britain, that is, the initial purchase of forty-five Jaguars and the assembly in India of another forty such aircraft, but not the licensed production of the remaining sixty-five Jaguars in India.

In the 1980s, there has been confusion over India's aircraft procurement strategy that has yet to be resolved. Doubts about whether to manufacture the Jaguar under license in India led to consideration of doing this with some of the Mirage-2000s instead. But neither solved the problem of Pakistan's ability to deploy the F-16s immediately. Under these circumstances, reasons advanced for cancelling the third part of the Jaguar agreement and to proceed with the purchase of the Mirages or even to coproduce them in India later appeared less relevant, since neither adjustment offset the Pakistani F-16s already deployed. Subsequently realizing that the Mirages would not be available for deployment until about 1987, India proceeded to purchase forty MiG-29s from the Soviet Union in 1984 to counter the American delivery of the F-16s to Pakistan.

The clearest and most long-term decision is illustrated perhaps in the Indian decision to manufacture the MiG-27 Flogger-D under licensed agreement with the Soviet Union. The MiG-27 is basically an advanced version of the MiG-23 Flogger, which India purchased in small numbers in its $1.6 billion arms package deal with the Soviet Union in May 1980. The MiG-27 would be the logical successor to the manufacture of the MiG-21 in India, especially since much of the necessary infrastructure for the manufacture of MiG aircraft already exists in India. Other important arguments advanced for continuing the coproduction of Soviet aircraft in India include the fact that their aircraft are technologically less complex and do not change rapidly in design from one generation to another.

Naval Defense Capabilities

The defense programs of the navy have been comparatively smaller than those of the other two services. Until 1965, the navy received less than 5 percent of the annual defense budget, while its vessels and equipment, of World War Two vintage and mainly purchased secondhand from the British, deteriorated. Thereafter, a steady rearmament program was begun. The navy's share of the defense budget had increased to about 10 percent by the early 1970s, and this has been maintained since then.[29]

The naval program over the last fifteen to eighteen years has emphasized purchases of a variety of Soviet naval vessels, which were added to the older vessels purchased from the British. Weapons and equipment for the naval air force were also primarily purchased from other countries, including Britain, France, the Soviet Union and the United States.

Thus far, only two major vessels have been licensed for production in India: the Leander frigates in collaboration with Vickers and Yarrow of Britain and the proposed SSK-1500 Type 209 subma-

[29] See Thomas, *Defence of India*, pp. 207–217; and *Report: 1979–80*, Ministry of Defence, pp. 17–20.

rines with Howaldtswerke Kiel of West Germany.[30] The manufacture of the frigates, which began in 1967, has now been indigenized, further developed, and adapted, incorporating more recent technological advances. For the naval air force, helicopter squadrons and jet trainers are obtained through licensed or indigenous production. These are the Alouette-III helicopter, made in collaboration with Sud-Aviation of France, and the Indian-made HJT-16 Kiran jet trainer. Procurement through domestic production has been made possible in these cases only because the craft are ordered in sufficient numbers by the air force and army as well. For the rest, indigenous manufacture for the navy by the three defense public sector enterprises, Mazagon Docks, Garden Reach Shipbuilders and Engineers, and Goa Shipyards has consisted of smaller and less sophisticated vessels. (See table 7.7.)

Naval procurement policies are conditioned by the small number of vessels that is usually needed by the navy, which requires that certain types of vessels be virtually custom built if they are produced indigenously or by license. For example, the frigate program at Mazagon Docks had by 1985 produced eight frigates (six Leander-class and two Godavari-class) in about eighteen years. The proposed submarine construction program with HDW is at present scheduled to build only four submarines over the next five to seven years. Under these circumstances, Indian efforts to design and develop combat vessels imply larger production costs and probably longer delays and lower quality than if production is undertaken in collaboration with an established overseas manufacturer.

However, it is important to note that apart from the production and fitting of armaments to battleships, the technology of battleship construction is not fundamentally different from that of commercial vessels. Moreover, the technology of armaments such as antiaircraft guns and surface-to-air and surface-to-surface missiles is basically similar to that of weapons being produced for the army and the air force. Therefore, a coordinated military-civilian shipbuilding program and a coordinated and more integrated armaments program for all three services should allow Indian defense planners to minimize the technological risks, time delays, and excessive unit

[30] For a discussion of these programs and agreements, see *Economic Times* (Bombay), December 7, 1979; Bharat Kanrad, "Substandard Subs for the Navy," *Hindustan Times*, November 21, 1982; Ashley J. Tellis, "Limitations of the Sea Harrier," *Times of India*, September 7, 1982; and Usha Rai, "Sentinel of the Deep Sea," *Times of India*, December 19, 1982.

TABLE 7.7
Navy Weapons Procurement, 1950–1985

Indigenous Production

seaward defense boats, survey vessels and crafts, landing craft, and torpedo recovery vessels

Licensed Production

Leander/Godavari frigates (Britain)[a]

SK-1500 Type 209 submarines (West Germany)[b]

Alouette-III/Cheetah helicopters (France)[a]

Overseas Purchases

Whitby frigates; Fiji cruisers; one Majestic aircraft carrier; Ton coastal vessels; Ham minesweepers (Britain)

Osa-I and -II missile boats; Petya-II frigates; F-1 submarines; Kashin-II destroyers; Nanuchka corvettes; Polnocny landing craft; Natya minesweepers (Soviet Union)

Seahawk fighter aircraft (Britain)[d]

Harrier VSTOL fighter aircraft (Britain)

Alize helicopters (France)

Ka-27 Helix ASW helicopters (Soviet Union)

Super Constellation maritime reconnaissance aircraft (United States)[d]

Tu-142m maritime reconnaissance aircraft (Soviet Union)

Sea King helicopter (Britain)[d]

Seacat SAM and Styx SSM (Britain)

[a] Weapon now entirely or almost entirely indigenized

[b] Licensed production of weapon had not yet commenced as of 1984

[c] Weapon is also to be produced under license in India.

[d] Weapon phased out or is being phased out.

production costs of designing and developing warships within India.

The navy's role in the future defense of India is expected to grow with the recent changes regarding maritime boundaries under the various international laws of the sea conferences. India's maritime boundaries now include much of the continental shelf around the peninsula and are extended by the Indian island possessions in the Bay of Bengal and the Arabian Sea. Concurrently, India's seaborne trade has been expanding rapidly over the last fifteen years. Both these two trends suggest growth in the scope of shipbuilding activities in India, both commercial and military. The indigenous construction of warships is likely to increase and become commercially more viable over the next two decades.

Nuclear and Space Programs

As discussed above, India's nuclear weapons and delivery capabilities must be inferred from its peaceful nuclear energy and space programs. Officially, there are no military programs in these areas; this is likely to remain the official posture as long as parallel military-related (or military-convertible) programs can be continued in the civilian sector. Material for bombs will always be available to India through either the uranium enrichment process or the reprocessing of spent fuel unless some technological development occurs that would separate a peaceful nuclear program from a military one. The rocket launch vehicles in India's peaceful space program, similarly, constitute ballistic missile delivery capability, and the weight of satellites for civilian purposes that a peaceful rocket may carry also indicate the payload that a missile might bear for military purposes.

Of course, discontinuation of the nuclear energy program due to commercial reasons, hazardous environmental conditions, or the availability of cheaper alternative sources of energy would require India to take up the path to nuclear weapons capability more explicitly. Nuclear programs would then have to be justified on the grounds of military necessity rather than economic viability in the civilian sector. This was, for example, the policy approach in China under Mao Zedung.

The situation is somewhat different in the space program. While there are alternative sources of energy available—and perhaps more desirable—there are no "alternatives" to space programs. For instance, there are no substitutes for the launching of satellites for meteorological and telecommunications purposes that are considered to be necessary for weather forecasting and national radio and television broadcasting. The choice here is between forgoing or accepting technological progress in this area. Even if the space program is relatively expensive and even if these services are available through satellites launched by other countries, the need for self-sufficiency in the long run may be a necessary political policy that is unlikely to change in the future. In other words, although serious questions may be raised regarding the motivations underlying the "peaceful" nuclear energy program, such questionable motivations would be much more difficult to advance in the case of the "peaceful" space program.

Technological and economic self-sufficiency were the key words underlying the atomic energy program as mapped out by the Indian nuclear scientist and first director of the Department of Atomic Energy, Homi J. Bhabha, soon after Independence in 1947. Bhabha's long-term plan—strongly endorsed by Prime Minister Nehru—envisaged three stages of development. The first stage was expected to establish heavy water natural uranium-fueled nuclear reactors based on the Canadian (CANDU) design. Heavy water reactor technology was preferred to the U.S.-designed light water reactor technology for three reasons: HWR systems did not require enriched uranium; they were perceived to be more efficient than LWR systems; and because HWRS also produced plutonium, the fissile weapons-grade material, as a byproduct.[31] The second stage was expected to establish uranium-fueled fast breeder reactors. Since such technology involved the generation of a "plutonium economy," the nuclear weapons option would be maintained simultaneously at a much higher level of potential capability. The third stage was expected to establish primarily thorium-fueled fast breeder reactors that would lead to a self-sustaining thorium-uranium fuel cycle. Since India has vast resources of thorium, the nuclear energy program was expected to become commercially viable after the year 2000.

Although the Bhabha plan (later elaborated by the next director, Vikram Sarabhai) did not recommend light water reactors, successful negotiations with the United States led to an agreement in 1963 for the construction and installation of two light water reactors at Tarapur (TAPP-I and -II) in Bombay. The American decision to cooperate was the outcome of Eisenhower's "Atoms for Peace" proposal in 1958. This proposal assumed that through such collaborative ventures that made the benefits of nuclear technology accessible, countries without nuclear weapons could be persuaded not to acquire them. In 1969, the American-aided Tarapur nuclear power plant was the first to be commissioned with a capacity of 420 megawatts of electricity (MWe).[32] The two light water reactors were provided to India on a turnkey basis by the General Electric Company of the United States. Indian industry and labor formed 30 percent of the project.

The heavy water reactors program began with the construction of the Rajasthan Atomic Power Plant based on two heavy water units of 220 MWe each supplied by the Atomic Energy Commission of Canada (RAPP-I and -II). (See Table 5.3.) The Rajasthan station has been operating since 1973 and was built with 70 percent participation and funding provided by Indian industry and labor. Since then, two heavy water reactors with capacities of 235 MWe each

[31] See Satish Jha, "How Not to Obtain Nuclear Energy," *Indian Express*, November 4, 1983.

[32] See *Nuclear News* 24, no. 2 (February 1981). For an official assessment, see the *Estimates Committee*, 1977–78, 16th Report, Ministry of Energy, 6th Lok Sabha, pp. 216–41. See also Raju G. C. Thomas, "Energy Politics and Indian Security," p. 41.

have been installed at Kalpakkam, a suburb of Madras (MAPP-I and -II), and two additional and similar reactors—with the same electricity-generating capacity—are being constructed at Narora in Uttar Pradesh (NAPP-I and -II). Both the Kalpakkam and Narora atomic power plants have been indigenously designed and constructed. When all the units at Tarapur, Rajasthan, Kalpakkam, and Narora are commissioned, India is expected to have a total nuclear electricity generating capacity of 1,800 MWe.[33] Meanwhile, there are plans to have ten Narora-type 235 MWe reactors in operation by the early 1990s, with twelve additional 500-MWe reactors on line by the year 2000. Should these various goals be attained, there will be as much as 10,000 MWe of nuclear electric power in the power grids of India by the turn of the century, almost all of which will have been developed exclusively by Indian technology, industry, and labor.

Despite these optimistic projections, the actual results have been less encouraging and have begun to call into question the economic viability of nuclear energy in India and the prospect of avoiding external dependence in the future. Since the Indian test of an atomic device in May 1974, Canada has withdrawn from further technical and material participation in India's nuclear energy program. The main setback has been the denial of heavy water for the CANDU-design power plants. Meanwhile, India's efforts to acquire heavy water from its own plants have proved to be a dismal failure. The heavy water plants built in India have functioned at about 7 percent of capacity and only three of the five plants were operative in 1983.[34] Leakages and loss of heavy water reported in the operating plants have led to the idling at times of the Rajasthan and Kalpakkan plants. In order to keep the Rajasthan plant in operation, India turned to the Soviet Union for the import of heavy water and this external dependence appears likely to persist for several years to come.

These setbacks led to the constitution of an ad hoc investigative committee appointed by Prime Minister Indira Gandhi and headed by N. B. Prasad, a former energy secretary in the central government. The committee's investigation, submitted in 1983, questioned the underlying basis of the Bhabha-Sarabhai long-term nuclear plan based on heavy water reactors. According to the Prasad Re-

[33] Figures derived from the *Performance Budget of the Department of Atomic Energy, 1980–81*, pp. 5–6. See also *Report: 1979–80* (New Delhi: Dept. of Atomic Energy, Government of India, 1980), pp. 4 and 7.

[34] Satish Jha, "How Not to Obtain Nuclear Energy," *Indian Express*, November 4, 1983.

port, if required quantities of heavy water cannot be produced, "the entire nuclear programme is in jeopardy, and a reassessment of our power programme becomes necessary."[35] The report even suggested the scrapping of the existing nuclear network to be replaced by a light water reactor program. It noted that in 1961, when the Indian decision was made to build the Rajasthan power plant, only 36 out of the existing 529 power reactors in the world were based on the heavy water design. The committee concluded that the essential objective of the Bhabha-Sarabhai plan—to insulate India from external dependence and pressures—was not being realized and warned that the heavy water nuclear reactors presently under operation or construction could soon become paralyzed.

The Prasad Report appeared to indicate the need to switch to light water reactors, with possibly the import of a 440-MWe reactor from the Soviet Union that the Soviets had reportedly offered in 1982.[36] From a political standpoint, the "Russian Option" seemed more desirable because of prevailing strategic alignments and international energy politics. The Soviet Union was already one of India's major suppliers of weapons and oil. However, reliance on the Soviet Union for an initial supply of light water reactors may face some of the same problems that India has encountered with the United States. Following the passage of the Nuclear Non-Proliferation Act in the United States in 1978, India has been unable to ensure the continuous supply of enriched uranium despite the existence of an international contract between the two countries for such supply until 1993. After considerable protests and negotiations, the supply of enriched uranium was eventually obtained from France (with Washington's endorsement) so that the U.S. government could comply with the new legislation passed by Congress. Further complications occurred thereafter with breakdowns in equipment at the Tarapur power plant, causing leakages of radioactive materials. Failure once again to obtain spare parts from the United States has compelled India to turn to West Germany for the necessary equipment.[37]

The Indian nuclear program in the early 1980s was clearly in disarray. The completion of plants at Kalpakkam and Narora was behind schedule and the successful completion of future heavy water plants in doubt. The acute political vulnerability of external de-

[35] *Statesman*, October 13, 1983.
[36] Ibid.
[37] Ibid.

TABLE 7.8
Growth of Manpower in the Departments of Atomic Energy and Space

	1979–80		1980–81		1981–82	
	Atomic Energy	*Space*	*Atomic Energy*	*Space*	*Atomic Energy*	*Space*
Scientific and Technical	17,037	6,193	18,021	6,863	19,158	7,136
Auxiliary and Administrative	11,041	3,422	11,651	3,903	12,498	4,172

SOURCE Adapted from the *Report· 1981–82,* Department of Atomic Energy, Government of India, p. 5; and the *Report: 1981–82,* Department of Space, Government of India, p. 37.

pendence was demonstrated in the noncooperation by Canada and the United States. Dependence on these two countries has only been replaced by alternative dependence on the Soviet Union, France, and West Germany. The problems here arise not merely from the questionable economic viability of nuclear energy programs, but from the ultimate political purpose of these programs. Nevertheless, the government of India is committed to its nuclear energy program no matter what the cost. Its justification is found in the need to maintain the nuclear weapons option and to employ the large pool of highly skilled scientists and engineers in this field who might otherwise migrate (see table 7.8).

Just as the growth of the Indian nuclear energy program is a reflection of India's nuclear weapons capability, the Indian space program is a reflection of India's missile or strategic delivery capability. According to the annual report of the Department of Space, the successful launching in July 1980 of India's first space launch vehicle (*SLV-3*) heralded a "major step forward not only for the Department of Space, but also for the nation as a whole."[38] However, the launch also demonstrated that India had acquired intermediate range ballistic missile (IRBM) delivery capability. When two major current projects reach fruition in the late 1980s—the Augmented Satellite Launch Vehicle (ASLV) and the Polar Satellite Launch Vehicle (PSLV)—India will be on the verge of acquiring intercontinental ballistic missile (ICBM) delivery capability.[39]

Programs of the Department of Space involve the development of satellite and space launch vehicle technologies. India took the first

[38] *Report: 1980–81* (New Delhi: Department of Space, Government of India, 1981), p. 5.

[39] *Statesman*, November 1, 1983.

modest step in space exploration on November 21, 1963, when it fired a 10-kg rocket within the atmosphere. The significant movement into outer space began with the launching of the 360-kg Indian-made satellite, *Aryabhata*, in April 1975 by a Soviet Intercosmos rocket carrier from a Soviet Cosmodrome. The satellite was intended for research in X-ray and gamma astronomy and aeronomy.[40] In June 1979, the first Indian-designed "Satellite for Earth Observation" (SEO-I), named *Bhaskara-I* and weighing 444 kg, was launched, again from a Soviet Cosmodrome. This was the first step toward developing an operational, remote-sensing satellite program for the retrieval and processing of remote-sensing data. A second satellite, SEO-II (*Bhaskara-II*), weighing 447 kg, was subsequently launched from a Soviet Cosmodrome in November 1981 and remained in operation for more than three years after its launch.

These projects have set the basis for long-term Indo-Soviet cooperation in space programs. In March 1983, India and the Soviet Union set up four joint groups for probing the earth from space, for meteorology and aeronomy, for astronomy, and for the optical observation of satellites. In the past, the two countries had established a joint working group for planning and preparing the launch of Soviet and Indian meteorological rockets from the test range in Thumba, Trivandrum. Also planned at the time were joint experiments in astronomy and the joint study of gamma radiation with the help of gamma telescopes installed on high altitude balloons to be launched in the area of the geomagnetic equator.

The second satellite system, the Indian National Satellite System (INSAT-I) is a multipurpose operational system for domestic telecommunications and meteorology with the capability for nationwide direct television broadcasting throughout India.[41] The first, INSAT-IA, was launched in early 1982 by a Delta expendible vehicle of the National Aeronautics and Space Administration (NASA) of the United States. This launch encountered a number of problems from the pre-launch stage and it finally broke down in space in September 1982.[42] With the first no longer operational, INSAT-IB was launched from the American space shuttle *Challenger* in August 1983. Consisting of two satellites, INSAT-IB differs from the other experimental satellites launched such as *Bhaskara*, *Rohini* and APPLE, in that it provides direct benefit to India's socioeconomic development through

[40] See report in the *Indian Express*, November 8, 1983.

[41] *Report: 1980–81*, Dept. of Space, p. 8.

[42] *Data India*, September 6–12, 1982.

television and radio broadcasting. Reservations had been made since then to launch INSAT-IC by an American space shuttle in 1986.[43] However, this launch is likely to be delayed due to the destruction of the *Challenger*.

The INSAT-I project is based on U.S. collaboration and assistance. The satellite itself was built by the Ford Aerospace and Communications Corporation, although it was designed by Indian scientists and engineers. Much of the electronic equipment in the satellite, for instance, was designed and fabricated at ISRO's Space Application Center. The electronics and avionics were principally provided by the government-owned public sector undertaking, the Electronics Corporation of India (ECIL). The critical portion of the civil works for INSAT-I Master Control Facility at Hasan in Karnataka was built by Indian engineers.

The Department of Space is now committed to the indigenous design, development, and production of second and subsequent generation INSAT satellites to eliminate the present dependence on the United States. This shift involves the indigenous development of the INSAT test satellites (Proto-INSAT) and the operational INSAT-II satellites.[44] However, the officially approved Space Research and Development profile for the decade 1980–1990 suggests that there would still be dependence on the United States for the launch vehicle to station these telecommunications satellites in space. Indeed, for future INSAT-II programs there exists a long-term arrangement with NASA for the provision of space launch vehicles and support services, at least through 1990.

Three other satellite development projects need to be mentioned. The first of India's experimental communications satellites was the Ariane Passenger Payload Experiment (APPLE). APPLE, weighing 670 kg, was first launched in 1981 and demonstrated Indian competence in building three-axis stabilized geostationary telecommunications satellites. Various tests for payload, vibration, and thermal balance were conducted at Toulose, France, and the satellite was then launched by the European Space Agency's *Ariane* rocket from a launching complex in French Guyana.

A second project is the Landsat earth station located at Secunderabad and operated by the National Remote Sensing Agency (NRSA) under the Department of Space. This station collects data from the American *Landsat-2* and *-3* satellites and the TIROS-N and NOAA-6 me-

[43] *Statesman*, October 16, 1983.

[44] *Report: 1980–81*, Dept. of Space, p. 9.

teorological satellites. Landsat satellite data acquired and processed at the earth station has been utilized for several survey projects of resources conducted by the NRSA. Data obtained here has been supplied to the International Locust Control project sponsored by the Food and Agricultural Organization of the United Nations, Bangladesh, Sri Lanka, West Germany, and the United States.[45] The data from meteorological satellites has been used to aid various flying missions by NSRA aircraft all over India.

The third project is the development of prolonged duration remote sensing satellites. The first in this series, called the Indian Remote Sensing Satellite (*IRS-1*), is to be built and launched around the late-1980s. The *IRS-1* is expected to weigh about a metric ton and will be in polar orbit about 900 kilometers above the earth's surface. The objective of this program is to acquire the ability to deploy spacecraft capable of remaining in orbit over three years and to beam information that will aid the exploration of natural resources on earth.

The *IRS-1* is expected to be launched by *Vikas*, a rocket with an indigenously built high-thrust liquid-fueled engine.[46] The super-precision engine for this rocket is being built using high alloy metals that can withstand high temperatures and pressure. The engine is capable of producing 4,000 horsepower with its turbo pump running at 9,600 rpm. The engine has already undergone tests at the Vikram Sarabhai Space Centre at Trivandrum and was undergoing further tests in France in 1983. The first launch is expected to occur in 1986 and thereafter more such rockets are expected to be produced in India every six to eleven months. ISRO engineers gained the needed technological expertise for developing the *Vikas* through their participation in the development of the *Viking* rocket engine of the European launch vehicle *Ariane*. The participation was made possible through an agreement between ISRO and the French firm Société Européenne de Propulsion (SEP). In exchange for the training provided by SEP, ISRO exported high pressure transducers built in India.

The most important project series in the Indian space program has been the indigenous development of the space launch vehicle, *SLV-3*, and the *Rohini* satellite. On July 18, 1980, India joined the exclusive club of nations capable of reaching beyond the earth's atmosphere when the *SLV-3* placed the 35-kg *Rohini Satellite-1 (RS-1)*

[45] *Report: 1980–81*, Dept. of Space, p. 14.

[46] See *Data India*, March 14–20, 1983.

in a 300 km by 900 km elliptical orbit.[47] Following this, a second *SLV-3* rocket successfully launched the 41-kg *Rohini* development satellite (*RS-D1*) in May 1981. A third launch of a development satellite was expected to follow soon. In addition to the continuation of the *SLV-3* launches, the Indian space program is well into the development of more powerful launches. Two of these are categorized as the Augmented Satellite Launch Vehicle (ASLV), capable of carrying a 150-kg payload (compared to the 35 kg of the *SLV-3* rocket); and the Polar Satellite Launch Vehicle (PSLV), capable of carrying a 600-kg payload.[48] Both of these two projects are expected to be completed by the end of this decade.

The Indian satellite program is much less dependent on external assistance than the space launch program. Satellites have been mainly designed, developed, and, with the exception of INSAT-IA and -IB, produced in India. Even INSAT-I, manufactured by the Ford Aerospace Company, was designed in India. On the other hand, the space launch program has depended thus far on assistance from other advanced industrialized nations. The *Aryabhata* and *Bhaskara* series of satellites were launched by the Soviet Union, the INSAT-I satellite series by the United States, and APPLE by the European Space Agency. The exception was the *Rohini* satellite series, which was launched by the Indian-made *SLV-3* rocket. However, the technology of the *SLV-3* is quite outdated.[49] Because of the uneven rates of progress in indigenous development between the satellite and space launch programs, the two have been deliberately delinked. This has enabled India to pursue one program without the other program slowing it down.

Whereas the peaceful nuclear energy program has faced several economic, technological, and international political setbacks, the peaceful space program has been able to continue without much hindrance despite the external dependence on the United States, the Soviet Union, France, West Germany, and Japan for assistance ranging from launch vehicles to a variety of components for the sat-

47 See *Report: 1980–81*, Dept. of Space, pp. 5 and 21–22.

48 *Times of India*, December 15, 1980.

49 According to Maurice Eisenstein, the *SLV-3* solid propellant rocket bears a striking resemblance to the Scout missile produced by LTV, a Texas-based aerospace company in the 1950s. By western standards, the *SLV-3* appears to be obsolete. If converted for launch on a ballistic trajectory, the Scout's range might be about 500 miles given a payload of 1,000 pounds. However, the Indian program expects to have higher payloads and orbits at latitudes of 23,000+ miles (see *The Washington Quarterly* [Summer 1982]: 112).

ellites.[50] However, a parallel indigenous space and satellite technological capability is being achieved rapidly alongside the various external collaborative efforts. There are apprehensions that international political pressures may extend to the space program in the future as the links between the space and nuclear programs become more apparent. This may happen if the nuclear energy program can no longer be justified for civilian purposes and the "dedicated path" toward military capability becomes more apparent. Under these circumstances, the cost factor in the nuclear and space programs will become irrelevant, and the need to acquire indigenous capability will become paramount—assuming of course that the maintenance of the nuclear weapons option is still perceived to be critical. Such a situation may lead to greater international suspicion of—and therefore perhaps less international cooperation in—India's space programs.

In general, technological growth advances both development and defense objectives. Because technological capabilities constitute a double-edged sword, the ultimate purpose of technological strategy in India has appeared blurred and often deliberately so. Capabilities in the nuclear and space programs are good examples of the dual purpose of technological growth. However, more conventional areas of technology—including the fields of aeronautics, shipbuilding, electronics, metallurgy, and mechanical engineering—have also been utilized to advance India's defense capabilities.

[50] Apart from the above external collaboration, there have been other collaborative efforts. The Satellite Telecommunications Project (STEP) was made possible through use of the Franco-German satellite "Symphonie." The Indian Space and Research Organization has collaborative agreements with the Centre Nationale Etudes Spatiales (CNES) and the Société Européenne de Propulsion (SEP) of France; the Deutsche Forschungs und Versuchsanstalt für Luft und Raumfahrt (DFLVR) of West Germany; and the Institute of Space and Aeronautical Science (ISAS) of Japan. See *Report. 1980–81*, Dept. of Space, pp. 43–47.

CHAPTER EIGHT

Security Perspectives and Prospects

THE ASSASSINATION of Prime Minister Indira Gandhi on October 31, 1984, and the overwhelming electoral victory of the Congress party led by her son, Rajiv Gandhi, at the end of the year, marked a watershed in the politics of India. Will the Congress government of Rajiv Gandhi continue with the policies of the past or provide new ideas and directions?

A review of the Indian political scene since Independence reveals the importance of prime ministerial leadership in the substance and direction of both domestic and foreign policies. Policies under the Nehru and Indira Gandhi administrations were responses to different domestic and international political conditions but were reflections as well of personal ideologies and political styles. In the fourteen years between the 1947–1948 Indo-Pakistani war and the 1962 Sino-Indian war, Nehru's tenure was one under which India experienced relative stability at home, avoided wars with its neighbors, and enjoyed enormous prestige as one of the leaders of the nonaligned movement. Nehru's ideology and style were basically socialist and idealistic. In retrospect, however, the 1962 war with China may be perceived by some analysts as the failure of Nehru's foreign and defense policies. Peace was achieved for fourteen years without an adequate Indian buildup of military capabilities that might have enabled India to negotiate with its neighbors from a position of strength.[1]

On the other hand, Indira Gandhi's rule occurred during a turbulent period of post-Independence history, characterized by political upheavals in the Indian states, increasing communal strife, a war with Pakistan, tribal insurgencies, a severe domestic economic crisis generated mainly by the international oil crisis, and a constitutional crisis arising from the suspension of the Indian democratic system for twenty-two months. Indira Gandhi's ideology swung from the continuation of Nehru's heavy socialism during the first eight years (1967–1975) to more freedom of opportunity for the

[1] See Raju G. C. Thomas, "Nonalignment and Indian Security: Nehru's Rationale and Legacy," *Journal of Strategic Studies* 2, no. 2 (September 1979):153–71.

private economic sector during the last six years (1975–1977, 1980–1984). Her foreign and defense policies moved India even closer to the Soviet Union than during the Nehru era, and then moved India back toward the center during her second administration that began in January 1980. Her political style was hard hitting and pragmatic in dealing with the crisis-ridden period in which she held power.

The beginning of the tenure of Prime Minister Rajiv Gandhi, therefore, raises questions regarding the future handling of India's internal and external security problems. There are aspects of the personality and political experience of the new prime minister as well as various changes in conditions at home and abroad in the mid-1980s that suggest minor and major shifts in India's policies—strategic, political, economic, and technological. Although part of the shift may be due to the change in political leadership, part may arise from what two authors have called the general shift in world concerns from "high politics," focused essentially on national security problems, to "low politics," concerned primarily with international economic and social issues.[2] The shift implies a lessening of Indian concern about East-West struggles and Great Power involvement in the subcontinent, and more attention to the progress in North-South negotiations, where political alignments suggest greater cooperation with India's traditional adversaries, Pakistan and China.

I. Review of Indian Security Policies

The Strategic Environment

There have been changes in India's strategic environment of varying degrees and importance at the global, regional, and nuclear levels. At the global level, the triangular relationships among the United States, China, and the Soviet Union underwent significant changes in early 1985. The first major economic and cultural agreement between the Soviet Union and China since the Sino-Soviet rift was followed immediately by an agreement between the United States and the Soviet Union to renew arms control negotiations.[3] Thus, although relations between Washington and Beijing continue to remain cordial and cooperative on a wide array of diplomatic, economic, and military-related areas, the implications of closer ties

[2] Robert O. Keohane and Joseph S. Nye, *Power and Interdependence: World Politics in Transition* (Boston: Little Brown, 1977), p. 24.

[3] *New York Times*, January 10 and 12, 1985

between Beijing and Moscow for Sino-American relations remain to be seen. A conceivable scenario over the next few years may involve closer Indo-American ties and the revival of the thaw in Indo-Pakistani relations. After some promising moves under the Janata government in the late 1970s, relations between India and Pakistan had begun to harden again after the Reagan administration's decision to allow Pakistan to purchase advanced American weapons to counter the Soviet invasion of Afghanistan.

Despite the Soviet military presence in Afghanistan and sporadic Indo-Pakistani efforts to normalize relations, perceptions of threats from each other still dominate the regional strategic assessments of India and Pakistan. From the Pakistani perspective, there are fears of potential Indo-Soviet military collusion to squeeze Pakistan out of existence. Moscow may want to set up Soviet satellite states in Baluchistan and "Pakhtunistan" to consolidate its hold over Afghanistan and to gain access to warm water ports on the Arabian Sea; India may want to reabsorb Sind and Pakistani Punjab into a greater India, thereby making the problems of Indo-Pakistani strife into a domestic issue of Hindu-Muslim relations. From the Indian perspective, Pakistan has not yet given up its claims to Kashmir and may be tempted to encourage the creation of a Sikh state of "Khalistan" in the Indian Punjab in order to make the Indian defense of Kashmir difficult.

With reference to the other part of the "traditional threat," Sino-Indian relations continue to show improvement, although a formal agreement on the border dispute that led to war in 1962 has yet to be reached. According to the optimistic Indian perspective, Sino-Indian issues appear less explosive than Indo-Pakistani disputes. There are, no doubt, fundamental differences in the ideology and political systems of India and China; but these have been accommodated in the past under the slogan of "peaceful coexistence" and may be similarly accommodated in the future. In retrospect, it could be argued that the territorial dispute that led to war in 1962 was the result of a series of misunderstandings and miscalculations of the intentions of the other.[4] At present, each side holds the territory it considers strategically important to itself: China holds the Aksai Chin plateau, through which a road was built to connect Tibet with the important Xinjang province; India holds the territory south of

[4] Allen S. Whiting, *The Chinese Calculus of Deterrence: India and Indochina* (Ann Arbor: University of Michigan Press, 1975).

the MacMahon line, which it considers critical for the defense of the Indian plains in Assam and West Bengal.

On the other hand, it is also important to note that these two major Asian powers, containing almost half the world's population and with fundamentally different socioeconomic and political systems, have more than a thousand miles of common mountain borders. The arms buildup continues in both India and China and, in particular, parallel strides in the development of India's potential nuclear weapons and delivery systems make these two states prospective nuclear rivals in the future. Therefore, according to the more pessimistic perspective, the 1962 Sino-Indian war must be seen both as a historic landmark and a reminder that relations between the two countries could swiftly deteriorate under a series of unforeseen circumstances.

On balance, threats emanating from within the subcontinent are still perceived to be greater than threats that may exist beyond the subcontinent. Although Indo-Pakistani relations have improved, for Pakistan, the Indian threat is perceived to be more serious than the Soviet threat, and for India, the Pakistani threat remains greater than the Chinese threat. Under these circumstances, proposals made at various times for the joint defense of the subcontinent become possible only when both India and Pakistan perceive the extraregional threat to be greater than the intraregional threat. In 1962, for instance, Pakistan refused to accept the Chinese threat to the subcontinent as greater than the Indian threat and strongly protested American efforts to arm India. In 1980, India refused to accept the Soviet threat to the subcontinent as greater than the Pakistani threat and strongly protested the American arming of Pakistan.

Although conditions under the traditional threat perimeter have shown favorable change, conditions under the extended strategic perimeter remain relatively the same. The arms buildup in Saudi Arabia, the war between Iran and Iraq, Pakistani military ties with conservative Arab states, and India's strong economic ties with this region continue to make developments here significant for Indian defense planning. The new strategic, political, and economic linkages between the subcontinent and the Persian Gulf mean that major crises in one region produce strong ripple effects in the other.

Similarly, in Southeast Asia, Vietnam's occupation of Kampuchea and the periodic military skirmishes on the Thai border continue to make this region of some concern for India. Unlike the Middle East, arms buildups and conflicts in Southeast Asia have fewer implications for Indian security; Indian interests in this region are more diplomatic and economic than military. As in the case of the Middle East, there are prospects for increasing India's trade and investments with the ASEAN countries. Conflicts in the region tend to reduce such prospects. Vietnamese military actions in Kampuchea and along the Kampuchean borders with Thailand are likely to provoke conflict with China as happened in 1978. The similarity between the 1978 Sino-Vietnamese conflict and the 1962 Sino-Indian war and the military ties of both India and Vietnam with the Soviet Union led in 1978 to a stall in the diplomatic mending process between India and China. Nor did India's recognition and support of the Vietnamese-backed Heng Samrin regime in Kampuchea help to improve ties between India and the United States. None of this, of course, suggests Indian military involvement in Southeast Asia but merely the possibility that India's global and regional political alignments may be affected by conflict in this region.

Soviet-American naval rivalry in the Indian Ocean remains the same, although the intensity of the issue among the littoral states of the Indian Ocean has declined since it first erupted more than a decade ago. Nevertheless, the naval presence of the two superpowers affects regional military balances and strategies in view of the ability of one or both of the superpowers to intervene swiftly in the conflicts of the region. In particular, this continues to have a bearing on the growth and strategy of the Indian navy. India's increasing naval reach into the Arabian Sea and the Bay of Bengal is offset by the ability of one or both superpowers to restrain Indian naval actions in these waters.

The implications of these global and regional relationships and the linkages between the two levels for India's strategic posture appear less clear. The success of the shift from a military posture of "sufficient defense" to meet the Sino-Pakistani threat to one of "limited deterrence" that would make the resort to force against India by its traditional adversaries costly and unacceptable may be demonstrated rather weakly by pointing out that India has not been engaged in a war since 1971. But this situation may also be a function of the changing strategic environment and diplomatic efforts by New Delhi to foster greater economic and cultural ties with Pakistan and China rather than the composition and growth of Indian military capabilities posing a credible deterrence.

Thus, the Afghanistan crisis has made it unavoidable that Pakistan drop its confrontational attitude toward India on Kashmir and other traditional Indo-Pakistani issues. On the other hand, the same crisis has also enabled Pakistan to upgrade the quality of its military capabilities substantially, posing a more credible threat to seize the disputed state of Kashmir by force in the future. As regards the northern threat, the border dispute between India and China has been virtually settled along the existing lines of control although no formal agreement has been signed.

The strategy of limited deterrence expressed in the continuous growth of Indian military capabilities is in a sense an expression of India's self-image as a regional power with strategic interests and concerns in the arc of the Indian Ocean region from the Middle East to Southeast Asia.[5] The growth in capabilities is also an autonomous movement that reflects India's economic and technological growth in the aerospace, automotive, electronics, shipbuilding, and other defense-related industries. However, both factors propel India's military capabilities forward and provide it with a deterrent posture against neighboring adversaries and extend its military reach further east, south, and west of the subcontinent.

Probably the most complicating factor in India's present defense posture is the nuclear question. The problem is not simply whether India should become a nuclear power in order to enhance its conventional deterrent posture but whether the acquisition of nuclear weapons is unavoidable because of the growth of nuclear weapons capability in China and the possible spread of such weapons in Pakistan and countries further west. Unlike the 1960s and 1970s, when the nuclear issue was always perceived as a matter of choice, the 1980s may make the acquisition of nuclear weapons appear a matter of necessity, should Pakistan deploy nuclear weapons. A nuclear India and Pakistan may not produce mutual nuclear deterrence on the subcontinent. Despite a joint declaration in December 1985 by Zia ul-Haq and Rajiv Gandhi not to conduct preemptive strikes against each other's nuclear installations, both sides may be compelled to adopt a nuclear-war fighting posture since neither side will be sure that the other will keep its promise under conditions of stress.

However, neither side has yet overtly embarked on a nuclear weapons program, though each remains on the brink of doing so. This strategy of brinksmanship in nuclear weapons capability may serve the interests of both India and Pakistan for the time being. But India, under the compulsion of responding to the growth of

[5] Richard L. Park and Stephen P. Cohen, *India: An Emergent Power?* (New York: Crane Russak, 1979).

Chinese nuclear weapons or of demonstrating its importance and role as a regional power, may provoke Pakistan into acquiring a nuclear weapons capability; or Pakistan, convinced that nuclear weapons are essential for its survival, may provoke India into taking steps to meet the Pakistani nuclear threat. Alternatively, decisions further west in Israel, Iraq, or Iran to become nuclear weapons states might trigger similar decisions on the subcontinent. Without detailing the various ramifications of nuclear proliferation it is clear that these circumstances would warrant a revised assessment of India's strategic doctrines and defense posture.

Internal Security

Problems of internal strife had taken a dramatic turn for the worse in 1984, culminating in the assassination of Indira Gandhi. In reexamining and reformulating India's security policies, the main priority for the new Congress administration of Rajiv Gandhi has been to come to grips with the growing internal violence to resolve political differences. Although the crisis in the Punjab—both the violent terrorist tactics of Sikh extremists and the government's armed response with the Indian army—received much deserved attention in India and abroad, the situation is symptomatic of a growing malaise in the country.

The relative number of incidents of Hindu-Muslim communal violence and the casualties involved have increased in recent years throughout the country. The earlier crisis in Assam in 1982, primarily between Assamese and Bengalis, was a struggle between both linguistic groups and religious groups, because most of the Bengali immigrants into Assam were Muslims from the neighboring state of Bangladesh. The civilian violence here had claimed a death toll mainly of Muslim Bengalis, that was higher than that in the Punjab. And again the army had to be deployed to end the violence. It is true that the several guerrilla wars conducted by the Nagas, Mizos, Gharos, and other tribal groups in the northeast sector of the country have subsided or are near resolution in recent years.[6] But the issues have not been completely resolved to the satisfaction of all the factions within these populations and the army continues to be deployed in these areas to maintain peace and order. Meanwhile, violence arising from industrial and economic issues continues to occur

[6] Rajiv Gandhi's government signed an agreement in 1985 with the leader of the Mizo guerrillas, Laldenga, to end the insurrection in Mizoram (*News India* [New York], November 15, 1985).

sporadically as before. However, the earlier scare arising from the armed insurrection of the "Naxalite" Communist movement, which first erupted in 1967 in West Bengal and then spread to Kerala and Andhra Pradesh, has dissipated.

The challenge facing the new administration is to identify the underlying causes of internal strife and to seek appropriate political measures to mitigate or resolve them. Of equal concern is determining the role of the Indian armed services and paramilitary forces in dealing with internal violence. Two basic questions may be posed at this point. First, has there been a tendency since the 1975 Emergency declaration to seek military solutions to what are basically political questions calling for political accommodations? Second, should the regular armed forces be used at all to maintain internal security?

Unlike dealing with external armed threats to the nation, the utility of force to resolve problems of internal armed violence would appear to be limited. Indeed, there may be no final military panaceas for the problems of internal violence. In the case of guerrilla warfare conducted by rebel tribals in Nagaland, Mizoram, Meghalaya, and Tripura, the subjugation of the insurrections through counterinsurgency measures may be viewed as only a temporary solution until political accommodations can be achieved. After all, the ability to conduct prolonged guerrilla warfare is dependent to a considerable extent on the sympathy and material support of the local populations. Successful guerrilla warfare implies substantial local support. Under these circumstances, the basic grievances of the alienated group need to be redressed if these problems are to be resolved on a more permanent basis.

This observation is probably even more valid in the case of the strategy of terrorism adopted by Sikh extremists in the Punjab. Unlike successful guerrilla warfare against government forces, which must receive prior support from the local population, the random hit-and-run tactics of the terrorist against usually innocent victims need no such support. The weapons required for such purposes are relatively simple and their targets are difficult to identify or defend. However, the success of this strategy lies not in the terrorist attacks, which are rejected by the population, but in the government's responses. Since the terrorists are unknown, the entire ethnic group becomes suspect. Constant search and seizure of members of the group steadily alienates large sections of the group from the government. The dilemma that Indira Gandhi's government faced in the Sikh terrorist strategy in the Punjab was whether to demonstrate through armed counterforce that terrorism does not pay or whether to let the situation drift and then hope that the majority of the Sikhs themselves, tired of the uncertainty and fear, would expose and eliminate the havens of the terrorists.

The decision to order the Indian army to attack the Golden Temple in Amritsar was, in retrospect, the wrong political choice. The action in a single stroke hurt the religious sensibilities of all Sikhs and politically alienated the majority of them. More Sikhs—both terrorists and innocents—died in the six months following the government's counterattacks against the terrorists operating from the temple than from the murders inflicted by the terrorists themselves. This was particularly true in the aftermath of the assassination of the prime minister, which, ironically, may have otherwise caused the sympathy of many Sikhs to swing away from the demands and actions of the Sikh extremists for an independent Khalistan toward reconciliation with the Indian government.

The subsiding of guerrilla warfare in the tribal northeast after almost thirty years of prolonged struggle reflects both the war-weariness of the guerrillas themselves and the successful implementation of various social and economic reforms. Short of conceding a separate and independent state as demanded by the guerrillas, a series of political concessions has diffused the several crises in this region. On the other hand, the failure in the Punjab was due in part to the government's inexperience in dealing with the strategy of terrorism. It may also have been due to Indira Gandhi's more hard-line approach in dealing with problems of security. Although this tactic produced spectacular success externally in dealing with the Bangladesh crisis in 1971, it could not resolve the internal Sikh crisis in the Punjab.

The second question regarding the use of the regular armed forces for the maintenance of internal security carries broader and more long-term significance. The military has argued that it constitutes one of the important unifying forces in a country that is constantly subject to intense fissiparous tendencies. Under these circumstances, the increasing use of the military "against its own people" undermines the very stability and unity of India. Other arguments against the deployment of the regular armed forces within the country include the need to avoid domestic politicization that could set the stage for a military coup and the need to maintain the military in battle-ready condition to meet external threats at all times.

The insulation of the regular armed forces from domestic politics and violence implies the need to further expand the paramilitary forces, most of which are under the Ministry of Home Affairs and some under the Ministry of Defence. Although this would be easy to do, such a policy also conjures up more authoritarian measures by the ruling government without a corresponding increase in internal security. Past deployment of the Indian army has invariably brought about peace and calm in riot-torn areas, but the past deployment of paramilitary forces and armed police, however, has sometimes provoked further violence. Moreover, an increase in the paramilitary forces would constitute a rival military establishment to the regular armed forces. Without the same discipline and code of conduct that is instilled in the Indian armed services, the paramilitary forces themselves could constitute a threat to internal unity and stability. On balance, this probable risk may be well worth taking in order to shield the regular armed services from domestic politics.

Finally, it is important to note that a major reason for the breakdown of internal law and order may perhaps be found in the increasing power of the central government (a trend that gained momentum especially under the administration of Indira Gandhi) and the conflict of interests arising from the demands of various religious, tribal, and linguistic groups at the state or local level for more autonomy. Therefore, the growth of centrally controlled paramilitary forces and the erosion of the powers of the state and city police forces may be perceived as a further attempt to reduce individual state autonomy. Far from resolving problems of internal security, this trend may in effect increase such problems. Ironically, as the level of violence at the regional level increases, there is a temptation to increase the control and power of the central government further. Although this federal reaction may resolve problems in the short run, long-term solutions to India's internal security issues may lie in providing greater autonomy to the states. Clearly, therefore, determining what forces and what levels of force are appropriate to deal with internal security will continue to remain thorny issues for post–Indira Gandhi governments over the years to come.

The Political Structure and Decision Making

In the December 1984 general elections, the Congress party, led by Rajiv Gandhi, won an unprecedented 79 percent of the total seats contested and received more than 49 percent of the total votes cast. There may be important political implications in this sweeping victory of Rajiv Gandhi that are unmatched by those of either his mother, Indira Gandhi, or his grandfather, Jawaharlal Nehru. The victory could be interpreted either as a clear mandate for new policies or as a rejection of opposition disarray and incompetence during a period of intense crisis. Whatever the reason, the Congress party's overwhelming majority in the 544-member Lok Sabha could also be used to introduce major constitutional changes that may have a bearing on the conduct of external and especially internal security policies.

The near eclipse of more nationally based opposition parties such as the Janata party, the Bharatiya Janata party and the Communist Party of India and the defeat of some prominent opposition parliamentarians such as S. Chandrashekar of the Janata party and A. B. Vajpayee of the BJP may further erode the influence of Lok Sabha debates in shaping security policies. In 1985, two leading opposition parties in the new Lok Sabha were regional parties from the southern Dravidian states—the Telugu Desam of Andhra Pradesh and the All-India Anna DMK of Tamil Nadu. The Communist Party-Marxist (CPM) was the only national party with a significant number of opposition seats. With this limited and regional form of opposition representation, the extent and quality of foreign, defense, and internal security debates in the Lok Sabha may suffer. No doubt, as during the Nehru and Indira Gandhi periods, Congress Members of Parliament will provide opposing viewpoints on the government's security policies. However, this is an inadequate substitute for the more formal and incisive debates between government and opposition.

Although the Congress party for the first time received almost half of the electoral votes (about 49 percent compared to between 35 to 48 percent in the seven earlier general elections), the December 1984 elections indicate that significant numbers of voters remain opposed or prefer alternatives to Congress rule. With much of this electorate unrepresented in Parliament and with major opposition groups and leaders unable to express their dissent through the parliamentary process, most of the normal democratic channels for releasing political frustrations in India appear to be closed. The danger to the Indian democratic process may arise not only from the ability of the Congress government in parliament to limit the democratic provisions of the Indian constitution in order to deal with internal security issues but also from the possible temptation of opposition groups to express their demands through violent demonstrations and protests in the streets, leading to the breakdown of law and order. On the other hand, the popularity and confidence of the new prime minister and the new era of economic prosperity may result in greater political stability at home and allow the new Congress government to become more sensitive to the voices of the weak opposition in Parliament.

More specifically, questions remain regarding the utility and efficacy of the existing security decision-making system. As we saw earlier, the new higher decision-making system that developed in the mid-1970s reduced the role and input of the armed services in the security decision-making process. The service chiefs had little or no participation in the Political Affairs Committee of the Cabinet, which considered problems of defense in the broader context of political and economic stability. Their powers in the Defence Planning Committee were inadequate because of the dominant representation of civil servants in the committee. In effect, as India's military capabilities have increased, the input of the military into the defense policy-making process has decreased, indicating a further tightening of civilian control.

There is, of course, nothing to indicate that the growing size and capabilities of the Indian armed services and their decreasing power in the security decision-making process are calculated actions on the part of the ruling politicians. But the trend is apparent and the temptation to draw a connection will remain. Given the overthrow of civilian governments in Pakistan, Bangladesh and much of the developing world in general, the need for caution in India is understandable, especially as the size and capabilities of the military grow. On the other hand, the military in India have claimed that they have neither the interest nor the ability to take over the government. The country—as well as the composition of the armed forces—is too vast and diverse for the military to plot, coordinate plans, and seize and control all the important cities and towns even if it so desired. Meanwhile, the inverse growth of military capabilities and military inputs into decision making could prove harmful for the development and execution of sound security policies.

Perhaps one way to increase military input into the security decision-making process is to accept the army's proposal for establishing a Chief of Defense Staff or equivalent system. As we have seen, the continuous buildup of Indian military capabilities and their projection on the high seas and further west and east will require a greater coordination of service roles and resource needs. The grounds for the army argument that the size alone of the Indian defense establishment calls for central command, control, and coordination are becoming apparent. At the same time, a Chief of Defense Staff system would also strengthen the negotiating powers of the military with the civilian authorities. The CDS, for instance, might be invited to sit in on the meetings of the PACC on a regular basis. On the other hand, fears of army domination of the other two services or of a military coup overthrowing civilian democratic control will remain and opposition to the CDS system will continue. These fears may be resolved over time as greater balance is achieved in the force levels of the three services and if the CDS system is set up as an integrated military-civilian establishment composed of personnel from the armed services as well as from the Ministries of Defence, Finance, Planning, External Affairs, and Home Affairs.

Another problem in the structure of the security decision-making system may need to be addressed in the future. Although policy questions on external and internal security are considered in the broader PACC, essentially they are managed and implemented separately by the Ministries of Defence and Home Affairs. The regular armed forces and the control of defense production and procurement are under the Ministry of Defence while most of the paramilitary forces are under the Ministry of Home Affairs. However, as we have seen, the armed forces have been increasingly used to deal with internal security problems, especially when paramilitary forces have proved to be ineffective. All of this indicates confusion and cross-purpose in the location of decision making, for the maintenance of internal security, especially in the control over the instruments of force to maintain internal law and order and in the level of force to be applied.

Thus, given the enormous internal security problems in India in recent years and the need to mobilize substantial armed forces to deal with them, the question is whether the Home Ministry, with its several other nonsecurity functions and responsibilities, is trained and equipped to deal with the problem of internal security. Routine law and order questions may be appropriately considered under the domain of Home Affairs, but when specialized military functions such as counterinsurgency, antiterrorist strategy, the control of large-scale communal violence or the prevention of industrial sabotage are called for, a more specialized department may become necessary. Moreover, in order to avoid the internal use and consequent politicization of the regular armed forces, the size of paramilitary forces and resources assigned to maintain internal security must be sufficient to handle such problems at all levels.

Should then the maintenance of both external and internal security be transferred to the Ministry of Defence, perhaps renamed the Ministry of National Security Affairs? From an organizational standpoint this would be in keeping with the newly perceived need to consider problems of external and internal security as equally important and interrelated. However, although such a change would allow more effective security decision making and policy implementation, there may be danger in concentrating all the armed forces of the country under one ministry, thereby producing conditions for potential military coups. This problem might be resolved by rigidly separating the two wings of the military between external and internal security forces and placing them under separate civilian ministers of state, both under a minister of national security affairs directly responsible to the cabinet. Alternatively, a separate wing for internal security affairs under a minister of state may be created within the Home Ministry. This was in fact done in October 1985 when Rajiv Gandhi appointed Arun Nehru as minister of state for internal security and law within the ministry of Home Affairs.[7]

The Military and Bureaucratic Politics

Issues regarding the strategic roles, force levels, and the allocation of annual budget resources among the three armed services continue to be debated in the 1980s. The last major war fought by the Indian armed forces was in 1971. Despite the strategic transformation in the arc of the Indian Ocean from the Red Sea to the Straits of Malacca—the arms buildup in the Middle East, the breakup of Pakistan, the Soviet military occupation of Afghanistan, superpower naval rivalry in the Indian Ocean, and the American military withdrawal from Southeast Asia—there have been no dramatic changes in the strategic roles and relative force levels of the Indian armed services. The debate regarding new force distributions, particularly to provide a more effective and balanced role for the Indian navy, had leveled off by the mid-1980s. Internal security issues had begun to overshadow external security concerns.

Perhaps the exception to this trend is the growth of the Indian air force. Since the Reagan administration chose to sell the advanced F-16 fighter planes to Pakistan in 1981, there have been several Indian purchases of aircraft from France and the Soviet Union. Although the acquisition of the Jaguar aircraft was essentially planned as a replacement for the aging British Canberra bombers and Hunter fighters, the decisions to acquire the Mirage-2000 from France and MiG-27s and MiG-29s from the Soviet Union will increase the relative role of air power in future Indo-Pakistani conflicts. To be sure,

[7] *Far Eastern Economic Review*, October 11, 1985.

neither the Mirage-2000 nor the MiG-29 aircraft may match some of the advanced technologies and performance capabilities of the F-16 aircraft. But the forty American F-16s Pakistan acquired were a reminder of its arming by the United States in the 1950s and therefore became the rationale for the further rebuilding of the Indian air force.

Indian overreaction to the American sale of the F-16s may be reflected toward the end of the 1980s in the relative redistribution of resources among the three armed services in favor of the air force and in its greater role in the defense of India. The ability of the air force to continue to receive greater attention in resource allocations will, however, be limited by two factors. First, the Indian navy will still claim a greater share of the defense budget allocations, while the entrenched army bureaucracy will also argue its case on the basis of its labor-intensive defense programs. Second, the heavy cost of aircraft and the potential for enormous financial losses either in battle or through flying accidents may suggest to Indian defense planners that future acquisitions should be based on more carefully planned long-term needs and not on short-term reactions to Pakistani acquisitions alone. In general, given the ambiguous nature of the present strategic environment, various opposing arguments on service roles and strategic planning will remain difficult to resolve; the tendency over the rest of the 1980s may be to maintain the relative status quo as much as possible until the patterns of external threats are more clearly demonstrated.

Defense and the Economy

Whatever the pros and cons of the defense-versus-development debate, the empirical evidence since 1975 indicates little cause for alarm. India's annual economic growth rate has averaged nearly 6 percent of the GNP between 1975 and 1985, while defense spending has been maintained at a steady rate of about 3.5 percent of the GNP. Inflation during this period has averaged less than 10 percent annually. The higher annual GNP growth rate implies much greater absolute levels of defense spending than during the previous fifteen years. Indeed, this was one of the major arguments of those who advocated low and controlled levels of defense spending, namely, that low defense spending and higher economic growth rates would, in the long run, be more beneficial for the defense buildup than high defense spending and lower economic growth rates.

Nevertheless, it is important to note that the question whether there exists an inverse correlation between defense spending and economic growth rates is not necessarily resolved. The consistently high Indian economic growth rate since 1975 is essentially the result of good monsoon rains and harvests as well as the new policies of both the Congress and Janata governments to encourage and increase the level of private investment. Additionally, long-term defense planning that permits military modernization programs and new defense projects to be spread over several years has helped avoid sudden and severe shocks to the Indian economy. This became possible because of the relatively long period of peace on the subcontinent. The exception to this was during the early part of the 1980s, after the Reagan administration's decision in 1981 to allow Pakistan to acquire $2.5 billion worth of arms, which provoked India to acquire more arms than it might have normally. Moreover, much of the new arms procurement involved the heavy expenditure of foreign exchange, especially for the Mirage-2000 aircraft. But without a major war in sight, the terms of acquisition and payment could be negotiated on more favorable terms, thus avoiding any adverse consequences for the economy that might have occurred if payment had had to be immediate.

The problem of foreign exchange depletion because of overseas arms purchases is likely to lessen as India enters the arms export market in the future. Until now, several public sector undertakings under the Ministry of Defence, such as Bharat Electronics Ltd., Bharat Earthmovers Ltd., Mazagon Docks Ltd., Garden Reach Shipbuilders and Engineers, and Praga Tools Ltd., have exported mainly civilian goods. BEL, for instance, has exported high-frequency communications equipment to Iraq and is considering setting up a joint industrial venture in Tanzania.[8] BEML has exported earthmoving equipment to countries such as Bulgaria, Bhutan, and Burma. The Defense-controlled shipyards have exported smaller civilian vessels and earned foreign exchange from dockyard repairs and services offered to foreign vessels in Indian ports.

In 1984, however, India began to explore seriously the arms export market for such items as automatic rifles and field artillery produced in the Indian ordnance factories. Most significant has been the decision of Hindustan Aeronautics Ltd. to seek the export of the Kiran Mk-2 jet trainer/light attack aircraft and the HTT-34 turboprop basic trainer. Meanwhile, HAL has entered into agreements to produce under license in India up to 150 of the DO-228 all-weather

[8] *Report: 1982–83* (New Delhi: Ministry of Defence, Government of India, 1983), p. 60.

jet trainer aircraft with Dornier and Messerschmitt-Boelkow-Blohm (MBB) of Germany.[9] MBB will also help HAL build its own antitank/transport helicopter both for use by the Indian army and for export. Similarly, the MiG-27 Flogger is also being produced under license in India in collaboration with the Soviet Union. But as in the case of the MiG-21, the export of Indian-made MiGs and parts is not permitted.

The spectacular growth of the private industrial sector in India following the easing of governmental licensing and controls in 1975 has continued into the mid-1980s. Government liberalization is also being extended to various areas of defense production—at least in the area of intermediate products—where greater private participation is being encouraged. Until recently, the Congress party's dogma was that all defense production must lie in the hands of the state on the grounds that private profiteering should not be encouraged in the area of arms procurement and national security. The new policy is expected to stimulate the private industrial sector as well as to bring about the optimization of defense production in terms of quality and cost.

The Technological Momentum and Military Capabilities

Under the Congress government of Rajiv Gandhi, we may expect further rapid advances in the level of India's scientific and technological capabilities. Indeed, Rajiv Gandhi's interest in and commitment to India's technological progress may be greater than that of his predecessor. The new prime minister perceives solutions to India's widespread poverty to lie in establishing a self-sustaining technological base. The increasing technological momentum will have significant spillover effects on the growth of India's conventional and potential nuclear weapons capabilities. As discussed above, the growth of Indian military capabilities is, to a certain extent, unrelated to Indian threat perceptions and tends to be an autonomous trend based on civilian technological growth. The situation is similar to that found in the United States, where advances in technology often define defense needs. In effect, India's extended strategic interests in the Middle East, Southeast Asia, and the Indian Ocean arise as much from India's growing military reach, based on an autonomous arms buildup, as from perceptions of indirect threats prevailing in distant theaters of war that may affect the subconti-

[9] *Aviation Week and Space Technology*, September 24, 1984, p. 24.

nent. Technological growth, therefore, has made India into a regional power that may rival China in the future.

The 1981 American decision on Pakistan temporarily threw India's planned weapons procurement strategies into a tailspin. Especially in the area of combat aircraft procurement, the virtual abandonment of indigenous development and procurement in the 1970s in favor of coproduction with an overseas manufacturer suddenly gave way to a flurry of overseas purchases. The emphasis on licensed or coproduction of weapons in India has become difficult to sustain, given the ability of Pakistan and countries in the Middle East to gain access to some of the most advanced weapons available in the West. This has been particularly true of air force procurement. In the case of naval arms procurement no such urgency exists, since countries east and west of India do not appear to have embarked on a major naval buildup. However, the problem here remains the small volume of naval arms procurement, which necessitates the construction of virtually custom-built combat vessels in India or continued emphasis on overseas purchases. Only the weapons needed by the army remain, for the most part, within the technological capabilities of India—although even here, Pakistani acquisition of TOW antitank missiles from the United States has led to similar Indian purchases from France and Sweden.

Undoubtedly, India's weapons procurement strategies need to be more closely related to the changing strategic environment and the growth of Indian technological capabilities. Strategic necessities should determine India's weapons requirements and India's technological strategy should then be responsive to such needs. It may seem optimistic in practice to expect such a logical approach, especially when the trend elsewhere in the world has been for weapons to chase technological capabilities irrespective of strategic necessities. Even the growth of India's scientific and technical personnel is not properly correlated to India's development needs. Many remain unemployed, underemployed, or migrate in search of better opportunities. On the other hand, in the key sectors of aeronautics, electronics, and computer technology there are still shortages of trained personnel. The need for the defense sector to absorb part of this large pool of experts may therefore become an end in itself, not unlike the employment of the Indian technical work force by the civilian public sector corporations in numbers that are larger than is needed. Such conditions lead to lethargy, delays, and inefficiency in planning and procurement that the defense sector cannot afford.

II. CONCLUSIONS

It is useful to conclude with a discussion of Indian security policy on two levels: a macrolevel that involves an examination of various external security perspectives; and a microlevel that involves an examination of various domestic policy responses. At the first level, the security relationships in Southern Asia may be viewed along three dimensions: east-west conflict issues, north-south economic negotiations, and internal-external security interactions. At the second level, the domestic factors that shape policy making may be viewed along two dimensions: civil-military relations and the interaction of economic and technological policies.

Security Perspectives

Much of this study focused on India's external security policy as it attempted to deal with politics at the regional and global levels. In particular, we have seen how the confrontations among the United States, the Soviet Union, and China led to attempts by all three powers to seek allies and support in South Asia. Conversely, the regional confrontations and wars between Pakistan and India and between India and China have led to a search for arms and military support by Islamabad and New Delhi from both west and east. The interaction of such regional and global politics consequently produced political alignments and military ties among Pakistan, China, and the United States on the one hand and between India and the Soviet Union on the other. We may consider this pattern of security relationships as part of the east-west conflict dimension.

As we have seen, since the mid-1970s there has been considerable concern in India about the increasing problems of internal security. The types of internal violence that threaten to undermine India's territorial integrity and political stability have included inter-religious, -linguistic and -caste violence, guerrilla warfare, and terrorist attacks. Most of these internal security problems have produced external involvement. For example, Hindu-Muslim rioting has invariably provoked Pakistani protests followed by Indian counterprotests that Pakistani reporting of the communal riots was exaggerated as well as inflammatory. Again, tribal guerrilla warfare in the Naga, Mizo, and Gharo hills was allegedly aided in the 1960s by China via Burma and by Pakistan from what was once East Pakistan. Similarly, Sikh extremists were alleged by India to have been given official or unofficial sanctuaries in Pakistan. There can be little doubt that most of India's internal security problems would be exploited by Pakistan or China in the case of war with either country. Likewise, India may exploit internal rebellions in Tibet in China, and in Baluchistan, Sind, and the Northwest Frontier Province in Pakistan. This is the internal-external conflict dimension of security problems.

There is a third, north-south dimension that may be of increasing importance in examining the patterns and trends of security relationships in the region. At the beginning of this chapter, the growing shift in world politics from "high politics" to "low politics," implying an increasing emphasis on economic and social issues rather than on military threats, was noted. This shift is also reflected in South Asia, where economic development issues have assumed greater importance than national security issues. To be sure, India and Pakistan are still concerned with military threats in and outside of the subcontinent. Internal security concerns not only remain in both countries but have now also seized the attention of Sri Lanka, where the Tamil separatist movement poses a problem. However, efforts in the 1980s to foster regional economic cooperation indicate that some of the older security concerns—the Kashmir dispute and the Himalayan border dispute—are becoming less critical. Growing economic cooperation between India and its two traditional rivals, Pakistan and China, does not imply that the older territorial disputes that led to war will be easily resolved. However, political accommodation on these issues is more likely to occur when economic cooperation and interdependence is established among these countries.

The north-south (or rich-poor) dimension suggests that India, Pakistan, and the other countries of South Asia as well as China all are developing countries seeking better trading and investment terms with the advanced industrialized countries. As such they have common interests in international economic negotiating forums, especially in the promotion of the New International Economic Order (NIEO). These newer economic concerns indicate a different set of regional alignments that may find all the countries of South Asia as well as China leaning toward the United States and the West in search of foreign investment and technological transfers. Both India and Pakistan follow a similar economic strategy, and their economic relations are primarily with the West. Especially in India, where socialist policies dominated economic planning in its first twenty-five years as a republic, greater encouragement has been given over the last ten years to private investment.

The new Indian economic policy suggests not so much a lowering of government investments in the massive and usually inefficient and unprofitable public sector corporations as much as the removal of government obstacles that discouraged private investments. Until recently, many foreign investors from the western world—especially America—were deterred by the Indian investment climate. Ironically, it was the declaration of the Emergency and the suspension of the democratic process in 1975 that first brought about change in India as part of enforcing discipline and efficiency in the public and private work force. However, the new economic policies continued under both the Janata and the Congress governments after the Emergency was lifted.

One of the significant characteristics of the present economic trend is the rate of domestic-foreign joint industrial ventures in the private sector that are sanctioned by the Indian government. For example, from 1981 to 1982, the number of new manufacturing licenses granted to the private sector increased by more than five times over the previous year.[10] Compared to negligible American participation in the past, when British, French, German, and Japanese firms provided most of the collaboration, U.S. companies have now taken the lead in establishing joint industrial ventures. Between 1980 and 1982, for instance, the United States provided 210 of the 915 joint ventures sanctioned by the Indian government; Britain placed second with 189. Meanwhile, since 1979 the United States has also replaced the Soviet Union as India's leading trading partner.

The emphasis on economic concerns is exemplified at the regional level by the formation of the South Asian Association for Regional Cooperation (SAARC). This regional organization was first proposed by the late President Ziaur Rahman of Bangladesh and was modeled on the Association of Southeast Asian Nations (ASEAN).[11] The SAARC proposal sought to mitigate the sources of conflict in the region by emphasizing those areas where cooperation is both possible and mutually beneficial. The Bangladesh proposal led to the meeting of the foreign ministers of the South Asian countries in Colombo in April 1981. Five areas of cooperation were identi-

[10] *Christian Science Monitor*, special supplement on India, July 30, 1982.

[11] For recent studies of SAARC, see Shrikant Paranjpe, *India and South Asia since 1971* (New Delhi: Radiant Publishers, 1985); and Shamsul Haq, "South Asian Regional Cooperation," *Third World Quarterly* 5, no. 3 (May-June 1984): 3-9. See also P.V.R. Chandrasekhara Rao, "Regional Cooperation in South Asia: Low Profile Affair in High Tension Area," paper presented to the Wisconsin Conference on South Asia, University of Wisconsin, Madison, November 1984.

fied: agriculture, rural development, telecommunications, meteorology, and health and population control. It was followed by further exploratory meetings in Khatmandu, Islamabad, and Dacca between 1981 and 1982. At the Khatmandu meeting more items were added to the list of areas where cooperation was seen to be feasible: transport, sports, arts and culture, postal services, and scientific and technical cooperation. In August 1983, the foreign ministers of the seven South Asian states—Bangladesh, Bhutan, India, the Maldives, Nepal, Pakistan, and Sri Lanka—met in New Delhi and agreed to create SAARC. Finally, in December 1985, SAARC was formally institutionalized at the meeting of the South Asian heads of state in Dacca, Bangladesh.[12]

In order to minimize the probability of failure at an early stage, the SAARC declaration emphasized cooperation in the economic, social, cultural, scientific, and technological fields rather than in political and security areas. One of the main problems of establishing economic cooperation was the dominance of India and the fear among the smaller states that India might have too much bargaining leverage over the others, especially if economic relationships were established on a bilateral basis. At the same time, there were misgivings among Indian officials that the SAARC would become a forum for the smaller states to gang up against India on many issues. But these misgivings appear to be have been dispelled at the Dacca conference in 1985, which was characterized by conciliatory positions and personal cordiality among the heads of state.

The emphasis on "low politics" and the focus on north-south issues, therefore, may well bring about greater cooperation among India, Pakistan, and China and between these countries and the West. This does not necessarily imply a deterioration in Indo-Soviet relations. India will continue to maintain strong military and economic ties with Moscow as long as the regional and global arms races continue. To put it in conceptual terms, the regional effects of east-west issues will be modifed by north-south issues. Thus, when the concerns of the South Asian states are perceived along east-west lines of conflict—as in the afermath of the Soviet invasion of Afghanistan—the problems of regional security will intensify along old patterns of security relationships. On the other hand, when the concerns of the South Asian states are perceived along north-south lines, there will be a greater easing of tensions in the region and the political relationships will be along a new pattern.

Although east-west conflict aggravates tensions in Southern Asia

[12] *New York Times*, December 9, 1985.

and north-south negotiations lower tensions, will the interaction of internal and external security problems heighten or lessen tensions in the region? Perhaps the preoccupation of the states of South Asia with internal security issues implies an unwillingness to embark on external military adventures. A war with Pakistan provoked by India during the Sikh crisis in the Punjab would make the defense of Kashmir and the Indian Punjab difficult if not impossible. Similarly, Pakistan would be hardly willing to provoke a war with India given its refugee burden in the volatile Pushtu-speaking areas of the Northwest Frontier Province and the sporadic but continuing nationalist movements in Baluchistan and Sind. Both India and Pakistan could be tempted to take advantage of each other's internal security problems to advance their territorial interests in Kashmir and elsewhere on the subcontinent, but for the time being, it seems unlikely that this will occur. In the multi-ethnic nation-systems of South Asia, all states must be acutely aware by now that the territorial integrity and independence of each is critical for the survival of all. Although India assisted in the breakup of Pakistan in 1971, there were misgivings even then about the support of the East Bengali nationalists seeking a separate state. It was recognized that such Bengali nationalism could spread to the Bengalis of India or could set an example for other ethnic, linguistic, and religious groups such as the Tamils, Telugus, Marathas, Kashmiris, and Sikhs who seek independent status from India.

It is unlikely, for example, that the government of India would deliberately seek to aid the Tamil separatists in Sri Lanka, although such allegations have been repeated several times by the Colombo government. The Indian government is aware that there was a strong Tamil separatist movement for a Dravidistan in India long before the movement began in Sri Lanka. Successful Tamil separatism in Sri Lanka could spill over into the state of Tamil Nadu in India. Similarly, if Pakistan were to openly assist Sikh separatism in the Indian Punjab, this might give encouragement to similar Sind or Baluch separatism in Pakistan. "Balkanization" is always a possibility in South Asia and its prevention requires the cooperation of all the states of the region. To this extent, the internal-external security dimension suggests greater cooperation rather than conflict among India and its neighbors.

Policy Responses

The nature of the relationship between the Indian civilian sector and the military sector determines not only the quality of security policy making in India but also the chances of maintaining long-term political stability. There have certainly been no military coups or countercoups in India. The military has remained apolitical. But there are complexities in the present nature of Indian politics that do not augur well for the future. First, as noted earlier, the increasing use of the armed forces, especially the army, for the maintenance of internal security has been accompanied by the decreasing role of the armed services in the higher levels of the security decision-making process. Second, the growth of paramilitary forces in India provides not merely a counterweight to the growth of the regular armed forces but also increases the instruments of force in the hands of the civilian government. Both trends suggest the development of conditions that could lead to the breakdown of the democratic political system in India.

In the first case, the increasing reliance on the Indian army to maintain internal law and order has already led to a serious breakdown of discipline and morale, demonstrated after the Indian army's assault on the Golden Temple. Although only Sikh mutinies and desertions occurred in the Indian army in the aftermath of the confrontation, these mutineers and deserters had to be shot or subdued by other units of the army, providing the basis for fear and uncertainty within the military establishment. The incidents of 1984 involving the army were the culmination of several years of complaints by the armed services that "the military should not be used against its own people." From the perspective of the armed services, their role is to deal with external enemies. At the same time, their increasing internal use is steadily eroding the image of the military as one of the primary domestic symbols of national unity.

On the other hand, from the perspective of the politicians, the armed services continue to be seen as a useful and effective tool for maintaining internal law and order. It might be simple for the government of India to increase the size of paramilitary forces for this purpose; but their deployment does not seem to have the same effect as the deployment of the armed services. At higher and more widespread levels of violence, the deployment of the army is seen by Indian politicians as almost unavoidable. However, such internal military deployment has also necessitated greater civilian control of the armed services. From this perhaps stems the tendency among the civilian leadership to reduce military input in the security decision-making bodies, even with reference to issues of external security, in that the same body—the Political Affairs Committee of the Cabinet—discusses both internal and external security issues. The net result increases both the politicization of the military establishment and their dissatisfaction.

The second dimension of relevance in examining policy responses is the manner in which economic and technological policies interact. Since the beginning of the roll-on five-year defense plans that were adopted in 1969, efforts were made to link India's economic plans more closely with defense plans. However, the development of civilian technology in India's economic plans is not necessarily geared to the needs of defense technology. Although there have been marginal increases in the allocations to defense research and development, much of this is intended to improve the "low tech" needs of the armed services—field guns, antitank nondetectable mines, new forms of ammunition, and a variety of electronic and instrument devices in the field of optical, electro-optical, and holographic instrumention. The major research and development that is taking place in the mid-1980s is the continuous development of the GTX-37-14U jet engine for future Indian-made fighters. For the rest, defense technology is primarily a spillover from civilian technological growth.

The basic problem lies in the fact that the Indian economy cannot afford the costs and risks of developing "high tech" defense items, while the development of civilian technology in the automotive, aeronautics, shipbuilding, and electronics areas does not usually meet the high standards of technology needed by the defense sector. The exceptions are the nuclear and space programs, which simultaneously provide India with energy and telecommunications systems as well as with nuclear weapons and delivery system capabilities, both at high levels of development. In other areas, civilian technological needs and defense technological needs do not always match. Thus the technological standards and quality of civilian automotive and electronics products are not—and need not be—of the standard demanded by the defense sector. On the other hand, the standards required by the defense sector may be economically prohibitive for a developing country such as India. Thus one of the basic needs of Indian economic and defense planning in the future will be to bring about a greater optimization in the development of civilian and defense technologies without jeopardizing Indian economic growth and stability.

It should be apparent that this analysis of security perspectives and policy responses does not suggest insurmountable obstacles. Despite periodic political upheavals, India remains one of the few bastions of stability in the Third World. There is virtually no other developing country that is faced with its complex internal amd external security problems. The country has, for the most part, successfully addressed itself to these issues under the enduring principles of democracy, secularism, nonalignment, and the maintenance of a mixed socialist-capitalist economy. There are lessons that may be learned by other countries from the Indian performance.

INDEX

Other titles in the

Studies in International and Strategic Affairs

CENTER FOR INTERNATIONAL AND STRATEGIC AFFAIRS
UNIVERSITY OF CALIFORNIA, LOS ANGELES

William C. Potter, editor, *Verification and SALT* (Westview Press, 1980).

Bennett Ramberg, *Destruction of Nuclear Energy Facilities in War: The Problem and Implications* (Lexington Books, 1980); revised and reissued as *Nuclear Power Plants as Weapons for the Enemy: An Unrecognized Military Peril* (University of California Press, 1984).

Paul Jabber, *Not by War Alone: Security and Arms Control in the Middle East* (University of California Press, 1981).

Roman Kolkowicz and Andrzej Korbonski, editors, *Soldiers, Peasants, and Bureaucrats* (Allen & Unwin, 1982).

William C. Potter, *Nuclear Power and Nonproliferation: An Interdisciplinary Perspective* (Oelgeschlager, Gunn, and Hain, 1982).

Steven L. Spiegel, editor, *The Middle East and the Western Alliance* (Allen & Unwin, 1982).

Dagobert L. Brito, Michael D. Intriligator, and Adele E. Wick, editors, *Strategies for Managing Nuclear Proliferation—Economic and Political Issues* (Lexington Books, 1983).

Bernard Brodie, Michael D. Intriligator, and Roman Kolkowicz, editors, *National Security and International Stability* (Oelgeschlager, Gunn, and Hain, 1983).

Raju G. C. Thomas, editor, *The Great Power Triangle and Asian Security* (Lexington Books, 1983).

R. D. Tschirgi, *The Politics of Indecision: Origins and Implications of American Involvement with the Palestine Problem* (Praeger, 1983).

Giacomo Luciani, editor, *The Mediterranean Region: Economic Interdependence and the Future of Society* (Croom Helm and St. Martin's Press, 1984).

Roman Kolkowicz and Neil Joeck, editors, *Arms Control and International Security* (Westview Press, 1984).

Jiri Valenta and William Potter, editors, *Soviet Decisionmaking for National Security* (Allen & Unwin, 1984).

Bennett Ramberg, *Global Nuclear Energy Risks: The Search for Preventive Medicine* (Westview Press, 1984).

William C. Potter, editor, *Verification and Arms Control* (Lexington Books, 1985).

Rodney Jones, Joseph Pilat, Cesare Merlini, and William C. Potter, editors,

The Nuclear Suppliers and Nonproliferation: Dilemmas and Policy Choices (Lexington Books, 1985).

Gerald J. Bender, James S. Coleman, and Richard L. Sklar, editors, *African Crisis Areas and U.S. Foreign Policy* (University of California Press, forthcoming).

Library of Congress Cataloging-in-Publication Data

Thomas, Raju G. C.
Indian security policy.

"Written under the auspices of the Center for International and Strategic Affairs, University of California, Los Angeles."
Includes index.
1. India—National security. 2. India—Military policy. I. Title.

UA840.T464 1986 355'.033054 86–12203
ISBN 0–691–07724–X (alk. paper)

Raju G. C. Thomas is Professor of Political Science at Marquette University. He is the author of *The Defence of India: A Budgetary Perspective of Strategy and Politics* (Macmillan).

GPSR Authorized Representative: Easy Access System Europe - Mustamäe tee 50, 10621 Tallinn, Estonia, gpsr.requests@easproject.com

www.ingramcontent.com/pod-product-compliance
Lightning Source LLC
LaVergne TN
LVHW011114300926
849160LV00008B/373